THE GENETIC APPROACH TO HUMAN DISEASE

VINCENT M. RICCARDI, M.D.

Robert J. Kleberg, Jr. Center for Human Genetics
Baylor College of Medicine
Houston, Texas

THE GENETIC APPROACH TO HUMAN DISEASE

New York

Oxford University Press

1977

Second printing, 1978

Library of Congress Catalogue Number: 76-56676
Printed in the United States of America

To Gabriel J. and Frances M. Riccardi

FOREWORD

Within the last two decades we have increased enormously our understanding of the fundamental processes of heredity. This new biological understanding can have its most immediate impact on the public through improved medical practice. The message of this book is that it is well within the purview of primary health care to utilize advances in genetic knowledge to aid the public, especially for prevention and early diagnosis of serious disease.

One of the chief purposes of medical genetics is to give information to couples on risks of disease in future children, and to put the risks in perspective for them, both in relation to random risks and to the prognosis of the child, if affected.

It is unfortunate that there is as yet no general public awareness of medical genetics such that the majority of those at risk will actively search for the information they need from the medical or paramedical professions. It is even more unfortunate that the medical profession, in particular family physicians, pediatricians, obstetricians, and internists, does not look upon genetic counseling as an essential part of the primary care, about which there is currently so much discourse.

Both of these defects can be remedied: (1) by better educating our young people in human biology and (2) by teaching medical genetics to medical students as part of primary medical care, not as part of basic science.

It is to these ends that the author of this readable book devotes himself. The "genetic approach to human disease" will become a reality when we have a better informed public that demands it, and a medical profession that relates genetic counseling to primary care.

Arthur Robinson, M.D.
Denver, Colorado
March 1977

PREFACE

I am often asked: "What is medical genetics? Is there something I can read to give me an overview of clinical genetics?" My constant answer of "no" to the second question and my many attempts to formally answer the first were two reasons for writing this book. The text is intended to show readers what the clinical application of genetic principles can accomplish. It should serve as a focal point around which other knowledge can be gathered, since I have tried to give readers a general understanding of an *approach* to health problems. Emphasis will be on fundamental principles; specific disorders will be considered mainly to exemplify the principles.

The book is also an elaborate argument that for clinical genetics to have its fullest impact, through prevention and patient management, the genetic approach to human disease must be applied at all levels of health care. Thus, its long-range purpose is to make the genetic approach part of the planning and delivery of routine health care. The immediate purpose is to inspire nongeneticists at all levels of health care to make use of genetic principles in their patient care activities. This book has also been written for students of medicine, nursing, and all other health care professions.

This effort largely reflects my experiences with the Colorado-Wyoming Regional Genetic Counseling Program. That program, originally directed by me and now in its fifth year of operation, provides clinical services, professional training, and public education to residents of Colorado, Wyoming, Nebraska, and Utah. It has given us an understanding of the factors involved in implementing genetic health care in local communities. A key feature of this implementation has been the training and education of health providers at all levels, and the book evolved from these efforts. It is organized so as to clarify the need for the genetic approach to health care and demonstrate how naturally that approach fits in with all types of health care provision.

Vincent M. Riccardi
Houston, March 1977

ACKNOWLEDGMENTS

My warmest thanks go to John W. Littlefield and Lewis B. Holmes for providing me the wherewithal to practice and teach clinical genetics, and to Arthur Robinson for making the Colorado-Wyoming Regional Genetic Counseling Program a reality. Financial support for the Program has been provided by the Department of Biophysics and Genetics of the University of Colorado Medical Center, the Colorado-Wyoming Regional Medical Program, the Henry J. Kaiser Family Foundation, the National Foundation–March of Dimes, the Division of Medical and Health Services, Department of Medical and Social Services for the State of Wyoming, and the Colorado State Health Department. A grant from the Book Program of the Commonwealth Fund provided support for the production of this book. Additional assistance was provided by the Southeastern Wisconsin Chapter of the National Foundation–March of Dimes, the Department of Preventive Medicine of the Medical College of Wisconsin, and the Department of Pediatrics of the Medical College of Wisconsin, and Milwaukee Children's Hospital. My special thanks to my wife Susan, and also to Cheryl Jirschele, Georgia Morris, David Peakman, Nancy Bergren, Cindy Bergmann, Margaret Butler, and Bekki Klemme.

CONTENTS

CHAPTER ONE

GENETICS AS A ROUTINE PART OF PRIMARY CARE

Introduction

The cardinal feature of genetics is predictability: given certain conditions, the likelihood of occurrence or recurrence of specific phenomena is predictable. Clinical genetics entails the reliable prediction of certain human disorders as a function of the lineage and/or biochemical makeup of any two parents or potential parents. The specialist in clinical genetics is trained to maximize the accuracy of such predictions. In this way, many disorders can be prevented in the first place. In addition, the clinical geneticist, by virtue of familiarity with rare disorders, is able to contribute to the management of patients already born with such disorders. But this approach to human disease can and should be applied by all providers of health care. The principle of predictability and its corollary of prevention fit into all phases of health care. For every instance of human pathology the question should be asked: "Will it happen again?"; and, whether yes or no, the affected person or at-risk family should be informed of the answer. It is the purpose of this book to make this fundamental question and an answer to it a routine part of health care at all levels, especially primary care.

A series of recent biomedical advances has made it feasible to apply this genetic approach on a widespread, routine basis:

1. The number of specially trained, clinical geneticists has increased substantially. Not only are there more of these specialists, but they are more accessible to other practitioners.
2. Improved diagnostic techniques permit more accurate diagnosis, carrier detection, and prenatal diagnosis. This is true for malformations, inborn metabolic errors, and chromosome abnormalities.
3. Widespread screening programs for a variety of biochemical disorders are technically feasible.
4. The resolution of chromosome structure for routine clinical practice has been improved.

5. Several techniques for prenatal diagnosis (x-ray, ultrasound, and amniocentesis) are now routinely available.

6. Techniques for specific and nonspecific modes of therapy of genetic disorders have been improved.

7. There is increased general awareness of both the practical implications of genetic principles and the associated moral, ethical, legal, and even political dilemmas.

8. Finally, there has been a relative increase in the morbidity caused by congenital and genetic disorders, especially for the younger age groups, due to decreased morbidity from other causes, particularly infections.

Clinical genetic principles can be applied by all clinicians. As the term is used in this book, a *clinician* is someone who is able and willing to *assist* a person or family toward improved health by *directly* altering the conditions contributing to the impairment. This broad definition extends beyond the physician, for whom the term has been traditionally reserved. The point is that persons in many professional and paraprofessional disciplines are literally clinicians. Moreover, it is incumbent on clinicians in all health care disciplines to recognize when health impairment is due, in part or in whole, to a genetic cause. All clinicians must be able to determine whether a given disorder is genetic, possibly genetic, or not genetic, and be able to share that information with the patient or family and refer them, if necessary, to specialists for further assistance. By the routine application of this approach, the clinician can establish a realistic basis for preventing genetic diseases. The purpose of this chapter is to demonstrate the "how" and "why" of such an approach to human disease.

The genetic approach is an attitude; it reflects a desire to assist patients and their families in avoiding the occurrence or recurrence of the disorder in question. As such, it is to be distinguished from genetic counseling, which is only one part of the whole. What characterizes the genetic approach is the clinician's involvement with family members other than the affected individual. Thus, for example, in working with a cystic fibrosis patient, it is important not only to minimize the effects of the disease in that patient, but also to ensure that siblings are checked for milder forms of the disease and that the parents understand the 25% recurrence risk in future offspring. Ensuring that these obligations are met is the responsibility of each clinician working with the family, even though he or she would not necessarily be able to carry out all the specific tasks. *Ensuring that a family with a genetic disorder receives genetic counseling is a primary care responsibility*, even though the actual genetic counseling may be carried out by a specialist in a secondary or tertiary care facility.

The Family Unit as the Focus of Health Care

A clinician dealing with a patient who has a genetic disorder is obligated to assist other family members, current and future, in minimizing the impact of the disorder on their lives. A corollary to this obligation is the recognition that, in certain instances, the focus of health care shifts from the individual patient to the family. The importance of the family unit as the focus of health care goes beyond counseling for recurrence risks. It emphasizes that the burden of the disorder and its management have their greatest significance in terms of the family as a whole. For example, if a young couple has two children with chronic and severe cases of cystic fibrosis, a third child, although healthy, might exceed the family's emotional and financial resources. Focusing on the family unit as a basis for decisionmaking is also a reasonable response to the increased clamor to make decisions for "the good of society." As elaborated below, society, through its agencies, assumes the burden, financial and otherwise, of caring for patients with genetic and congenital disorders. The benefit of a publicly supported genetics program should be considered in terms of the maximum number of family units that are kept intact and functioning optimally.

Relative Prevalence of Genetic and Congenital Disorders

Genetic and congenital disorders are important at all levels of health care because they can involve all body parts and functions and because they are so common (Table 1-1). Major congenital malformations, defined as those which interfere with normal livelihood and/or require surgery, occur in 2 to 4% of all newborns. The incidence is even higher when assessed at one year of age, since by that time previously occult problems, such as urinary tract anomalies and deafness, have become manifest. Congenital anomalies have become more important as a cause of death as infectious diseases have been curtailed. Up to age 14 years, congenital malformations are the third most common cause of death in both sexes. From ages 15 to 35, they are the ninth most common cause of death in both sexes. Chromosome abnormalities account for at least 20% of all spontaneous abortions, as well as a significant amount of infertility. The incidence of chromosome aberrations among liveborn infants is at least 0.5%. Diseases inherited as Mendelian traits, though often rare when taken singly, have a collective incidence of about 2%. Moreover, some of them are quite common in certain ethnic or national groups, e.g., hyperlipidemias, cystic fibrosis, alpha-1-antitrypsin deficiency, sickle-

TABLE 1-1

Frequency or Incidence of Genetic and Congenital Disorders[1]

1. Congenital malformations, general (genetic and nongenetic)[2]	
a. Major malformations	
At birth, 2 to 4% incidence	
By one year of age, 4 to 6% frequency	**4%**
b. Minor malformations, 10 to 12% frequency	
2. Genetic, chromosomal abnormalities	
a. Among spontaneous abortions, at least a 20% incidence	
b. In newborn populations	
General, 0.5% incidence	**0.5%**
Among perinatal mortalities, 7% frequency	
3. Genetic, single-gene disorders	
Estimated combined incidence, 2 to 3%	**2–3%**
About 1900 traits have been characterized	
4. Genetic, polygenic disorders	
a. Individually (e.g., cleft palate), about 0.1%	
b. Collectively, more than 1% incidence	
5. Combined genetic and congenital disorders	**6–7%**[2]

[1]For each type of genetic and congenital abnormality an incidence or frequency estimate is provided. The cumulative likelihood of having a congenital and/or genetic disorder thus comes to 6–7% (item 5).

[2]Includes polygenic malformations. Data are derived from refs. 2 through 8.

cell anemia, thalassemia, Tay-Sachs disease. Such polygenic malformations as spina bifida/anencephaly, cleft lip/palate, club foot, congenital heart disease, and pyloric stenosis have individual incidence rates of about 1:1000. And, since there are more than 10 such disorders, their combined incidence is between 1 and 2%. Other polygenic disorders, such as diabetes mellitus and essential hypertension, are even more frequent. It has been estimated that approximately 20% of all pediatric hospital patients have a genetic contribution to their clinical problems; the figures are somewhat less for adult in-patients and for out-patient services.

In a recent survey of discharge diagnoses at the Colorado General Hospital spanning four years, I found that at least 1.5% of the admissions were for eight representative types of genetic disorders (neurofibromotosis, cystic fibrosis, polycystic kidney disease, sickle-cell anemia, neural tube defects, congenital heart disease, cleft lip/palate, and all chromosome disorders). Less than 6.5% of the charts of patients identified with these disorders showed that the genetic approach to their diseases had been part of the final disposition. That is, neither the formal consultations nor the discharge summaries indicated that the genetic nature of the disorder or the specific recurrence risk was discussed with the patient or family. Since genetic and congenital diseases account for such a large proportion of hospital admissions, I would suggest that every hospital discharge protocol contain two questions to be answered as part of the discharge process: (a) *Is the disorder* genetic? (yes, no, or possibly); (b) *Was the family/patient told?* (yes, no, or deferred).

Genetic and Congenital Disorders as a Public Health Problem

The potential for prevention is one of the three reasons for considering genetic and congenital disorders a public health problem, the other two considerations being the commonness of such disorders and the relatively large number for which government agencies assume the financial and management responsibilities (Table 1-2). If we consider just one of these disorders, Down syndrome, their public health significance becomes apparent in terms of all three criteria. Down syndrome occurs in about one of every 1000 births, and the incidence rises with maternal age, so that by age 35 years the risk is 1 in 100 and by age 45 years the risk is 1 in 40. Thus, it is a relatively common disorder. Also, it is often preventable by means of prenatal diagnosis for high-risk mothers; for example, approximately 25% of the patients with Down syndrome are born to women aged 35 years or older. And finally, it has been estimated that $400 million is spent annually in the United States for the care of patients with Down syndrome (1); the major share of that expenditure is assumed by federal, state, and municipal government agencies.

Down syndrome is only one of many genetic disorders that readily fit these three criteria. It would be worthwhile for the government to subsidize voluntary preventive genetic programs, but such programs would be effective only if clinicians at all levels of health care made sure that the patients and their families were routinely counseled and thereby given the opportunity to make rational decisions. This is what we are getting at through the notion of the genetic approach to human disease.

Indications for Referral to a Genetic Counselor

The most general indication for a genetic evaluation or referral to a clinical geneticist is the query, "*Will it happen again?*" Will a given disorder show itself again, taking into account the personal,

TABLE 1-2
Public Health Nature of Genetic and Congenital Disorders

1. Genetic and congenital disorders are common
2. They are potentially preventable
3. Their expense is frequently borne by government agencies—federal, state, and municipal

genetic, and environmental background of the parents or potential parents? That is, can we predict whether the disorder will occur, considering what we know about the parents themselves (e.g., maternal age, consanguinity, health), their prior offspring and other relatives, and the possible environment of the developing fetus (e.g., teratogenic exposure)? The specific answer in a particular case is not the criterion for deciding whether there should be a genetic evaluation or referral; rather, it is the question as posed in the first place. Likewise, it is not important who originally asks the question—whether it be the family, physician, nurse, or social worker. Making sure that the question is asked, when it should be, and directly or indirectly providing for an answer is what matters. By doing these things the clinician implements the genetic approach to human disease. The answer itself entails a decision based on genetic principles. The process of reaching that decision and effectively communicating it to the family is what takes place in genetic counseling. Circumstances that raise the question "Will it happen again?" are summarized in Table 1-3. These circumstances and conditions are more often

TABLE 1-3
Indications for Referral to a Clinical Geneticist

1. Known or presumed congenital abnormalities
   a. Congenital malformations of any type
   b. Ambiguous genitalia, hermaphroditism
   c. Mental retardation (cause unknown)
   d. Fetal or parental exposure to environmental agents (drugs, irradiation, infections, and maternal factors)
2. Acknowledged familial disorders
3. Known inherited disorders
4. Metabolic, biochemical disorders (screening)
5. Known or suspected chromosome abnormalities
6. Multiple miscarriages, stillbirths
7. Infertility
8. Premarital counseling
9. Consanguinity, incest
10. Prenatal diagnosis
    a. Either parent a known "balanced carrier" for a chromosome abnormality
    b. Previous child with any kind of chromosome abnormality
    c. Mother 35 or older
    d. Inordinate parental concern or anxiety
    e. Both parents carriers for a specifically diagnosable metabolic or structural autosomal recessive disorder
    f. Either parent affected with a specifically diagnosable metabolic or structural dominant disorder
    g. Mother a known or presumed carrier for a serious X-linked recessive disorder
    h. Either parent having a first- or second-degree relative with a neural tube defect (e.g., spine bifida, anencephaly).

than not encountered in a primary-care setting. In the Colorado-Wyoming Regional Genetic Counseling Program (see Chapter 16), for example, more than 90% of physician referrals for genetic evaluation and counseling came directly from general practitioners, family practitioners, pediatricians, obstetricians, and internists in private practice; and many of the rest came indirectly from them through specialists, especially neurologists.

Congenital malformations of any sort both document an aberration of fetal development and set the stage for various adverse effects, not the least of which are emotional disturbances and domestic strife in the affected family. Clarification of the etiology and recurrence risk through genetic counseling not only has the consequences for the management of the patient but also helps the family to deal with the overall situation as realistically as possible. Thus, all families that have children with congenital malformations should be afforded the genetic approach and, ideally, genetic counseling. This would be true even if the only abnormality is developmental delay or mental retardation when the cause is not apparent. Anomalies involving the genitalia and gonads should be evaluated by an experienced genetics-oriented team as soon as the defects are recognized. In this way, optimal management, including the proper gender assignment, and recurrence risks can be established before there are irreversible undesirable consequences (i.e., incorrect gender roles, birth of a second affected child).

A number of *environmental agents* have an established or a suspected ability to disrupt fetal development. Respect for the potential of these agents can lead to prospective concern about congenital malformations and deliberate, informed decisionmaking on continuation of a pregnancy. Retrospective assignment of an environmental cause for malformations can also be an important factor in deciding about future offspring (e.g., in epileptic women being treated with diphenylhydantoin).

A disorder may be recognized to "run in the family," and this raises the question of heritability. A careful genetic evaluation is important in order to determine whether the familial clustering indicates a genetic etiology. Merely knowing that a disorder is *heritable* is not sufficient for counseling; the specific pattern must be considered for accurate decisionmaking. Almost everyone recognizes that hemophilia A is heritable, for example, but the implications of its X-linked recessive nature are not always appreciated. The unaffected brothers of an affected man are without risk of having either affected sons or carrier daughters. All the affected man's daughters, however, will be carriers, giving each a 50% risk for her sons to be affected.

Metabolic or biochemical disorders may be heritable, and for many of them prenatal diagnosis and carrier detection are available. For some of these disorders, such as phenylketonuria and sickle-cell anemia, detection of the disease or carrier state is afforded by large-scale screening programs. When a person is identified as being affected or a carrier, the clinician is then obligated to clarify for that person and other family members what this means in terms of prognosis and recurrence risks. *It is wrong to merely demonstrate the presence of a mutant gene without informing the bearer of its significance.* For example, when an infant is found to have phenylketonuria through newborn screening, then older siblings must be tested and the parents informed of the recurrence risk.

When a disorder is known or suspected to be *chromosomal,* genetic counseling is important to document the type of chromosome abnormality, arrange for other family members to be tested if appropriate, clarify the anticipated course of the disease, ensure that the recurrence risks are understood, and make amniocentesis available in subsequent pregnancies of at-risk persons. Chromosome abnormalities are being recognized as responsible for increasing numbers of congenital syndromes and other clinical problems (see Chapter 3). These include *multiple miscarriages* and some instances of apparent *infertility.* (There are, of course, other genetic etiologies for these two problems.)

Premarital genetic counseling is being sought with increasing frequency by young couples who would like to know their baseline risks for birth defects and whether they should have any special concerns. *Consanguinity* represents a special instance in which premarital genetic counseling can be particularly helpful. *Prenatal diagnosis* as a specific indication for genetic counseling will be dealt with in depth in Chapter 8.

Cited References

1. Rimoin DL: Manpower needs in human genetics. *Clin Res* 23:61, 1975.

2. Warkany J: *Congenital Malformations-Notes and Comments.* Chicago, Year Book Medical Publishers, 1971, pp 38–40.

3. Marden PM, Smith, Dw, McDonald MJ: Congenital anomalies in the newborn infant, including minor variations. *J Pediatr* 64:357, 1964.

4. Carr DH: Chromosomal abnormalities in clinical medicine. *Progr Med Genet* 6:1, 1969.

5. Lubs HA: Neonatal cytogenetic survey. in Wright SW, Crandall BF, Boyer L (eds): *Perspectives in Cytogenetics.* Springfield, Il, Charles C. Thomas, 1972, pp 297–304.

6. Miller OJ, Breg WR: Current Concepts in Genetics: Autosomal chromosome disorders and variations. *New Engl J Med* 294:596, 1976.

7. Day N, Holmes LB: The incidence of genetic disease in a university hospital population. *Amer J Hum Genet* 25:237, 1973.

8. McKusick VA: *Mendelian Inheritance In Man—Catalogs of Autosomal Dominant, Autosomal Recessive and X-linked Phenotypes* (4th ed.), Baltimore, Johns Hopkins University Press, 1975, pp 1–837.

General References

A. Bartolos M: Genetics in medical practice of the 1970s. *NY State J Med* 73:2763, 1973.

B. Clow CL, Fraser FC, Laberge C, et al: On the application of knowledge to the patient with genetic disease. *Progr Med Genet* 9:159, 1973.

C. Fraser FC: Current concepts in genetics: Genetics as a health-care service. *New Engl J Med* 295:486, 1976.

D. Kushnick T: When to refer to a geneticist. *JAMA* 235:623, 1976.

E. Trimble BK, Doughty JH: The amount of hereditary disease in human populations. *Ann Hum Genet* 38:199, 1974.

CHAPTER TWO

GENETIC CLASSIFICATION OF HUMAN DISEASE

Introduction

To be useful, a system for classifying disease should increase understanding of the diseases encompassed and thereby enhance their management. An effective genetic classification would also emphasize the predictability of recurrence risks. The genetic classification described here, which is similar to others (1), has been used for patient accounting and record-keeping, statistical analyses, and as a teaching device and format for patient-care decisions (2). Underlying its use is the belief that every patient or family can expect to be informed about the likelihood of recurrence of the disorder that brought them to seek professional attention. When the information is straightforward, assistance from a geneticist is not necessarily required. But if there is any uncertainty, or if backup is needed, a specially trained geneticist can be consulted.

Health care providers, by definition, assume responsibility for assisting patients or families already afflicted by disease. But at all levels they must also assume responsibility for initiating measures that facilitate prevention (i.e., minimize recurrence). Because the prevention of most, if not all, genetic and congenital disorders depends on a family being properly informed, emphasis must be placed on the process by which the relevant information is gathered, formulated, and delivered. In this chapter I demonstrate a scheme for making this kind of preventive information immediately accessible to all families in need of it.

The Genetic Classification

A simplified, general form of our genetic classification of human disease is shown in Table 2-1. The table also makes clear the main value of the classification: prediction of recurrence risk. Any given disorder can be considered to have a genetic, possibly genetic, or *nongenetic* etiology. Ideally, every disorder would be

TABLE 2-1
The Genetic Classification of Human Disease—a Guide to Predictability of Recurrence

A. Genetic	Recurrence risk defined—high or low
B. Possibly genetic	Recurrence risk unclear
C. Nongenetic	Recurrence defined—low

classified as either genetic or nongenetic, and the risks for recurrence always thereby defined. But many disorders cannot be neatly categorized in this dichotomous fashion. Those that fall into the "possibly genetic" category are rare, nonspecific, or heterogeneous. These limiting cases are important since the classification scheme depends on making the most specific diagnosis possible. The genetic or nongenetic nature of some disorders is implicit in the diagnosis. Cystic fibrosis is always a genetic, single-gene disorder; congenital rubella is a nongenetic, environmental disorder. For certain other disorders the decision to include a particular case in one class or another depends on the clinician's previous experience and/or the details of the family history. For example, an otherwise apparently normal male child with hydrocephalus due to stenosis of the aqueduct of Sylvius could have developed this condition because of congenital toxoplasmosis (nongenetic), the additive effect of multiple minor genes (polygenic), a single gene, X-linked recessive mechanism, or other factors. With a negative family history it may be impossible to specify etiology. However, if the child has no affected female relatives, but does have similarly affected brothers, maternal uncles, and male cousins through his mother's sisters, then the etiology can be considered genetic (X-linked recessive).

As mentioned, this whole approach depends on accurate, specific diagnosis. The two general kinds of diagnosis are descriptive and designative. The latter identifies a disorder in much the same way as a brand name identifies a specific commercial product. The former does no more than itemize one or more characteristic features. Unfortunately, diagnoses that are actually descriptive not infrequently evolve into "specific designations," and, conversely, a "specific designation" may come to stand for several different disorders that consequently need to be specified by qualifying terms. The distinction here is between terms like "Down syndrome" (designative) and "retinitis pigmentosa" (descriptive). Thus, a diagnostic term can have various uses, and its implied specificity may be misleading unless it is clarified by other means, such as an etiologic classification of disease. In this context, the classification presented here can assist clinicians in establishing the etiology and recurrence risks of a patient's disorder. Inclusion in a *genetic* subcategory defines the risk, which may be high or low depending on the specific diagnosis, the pedigree analysis, and the relationship of the affected person to the individual who has sought the counseling. Inclusion in a *possibly genetic* subcategory indicates that although we cannot quantify the risk, if it is present at all, neither can we say that there is no risk of recurrence. Inclusion in a *nongenetic* category indicates that there is no intrinsic or inherent risk for recurrence of a disorder beyond that for the general population.

There are three basic types of *genetic* disorders: chromosomal, single gene (Mendelian), and polygenic (see Table 2-2). Each involves a change in the genetic material at the cytologic, chemical, or functional level. In *chromosome* disorders (A1) there is either too much or too little genetic material; or it has been rearranged in such a way as to interfere with gamete production and function, or with early fetal development. In *single-gene* disorders (A2) the defect in the genetic material is on a much smaller—submicroscopic or macromolecular—level. An individual gene or a pair of genes that has been altered in terms of single molecules is the primary abnormality. *Polygenic* disorders (A3) reflect gene abnormalities that are manifest only when they accumulate in sufficient numbers to exert their additive effects and when the deleterious influence of (usually unspecified) environmental agents is present. Each of these types of disorders will be discussed in Chapters 3, 4, and 5, but several points should be made at this juncture.

We consider virtually all types of chromosome abnormality—whether of chromosome structure or number—more likely to recur among first- or second-degree relatives of affected individuals than in the general population. When the basic defect involves

TABLE 2-2
The Genetic Classification of Human Disease—Subcategories for Assigning Specific Types of Disorders

A. Genetic
- A1. Chromosomal
- A2. Single gene
- A3. Polygenic
- A4. Genetic mechanism not specified

B. Possibly genetic
- B1. Probably genetic per family history
- B2. Probably genetic per past experience
- B3. Uncertain because of no specific diagnosis
- B4. Uncertain because of heterogeneity

C. Nongenetic
- C1. Developmental
- C2. Environmental
- C3. Other, nongenetic disorder
- C4. Normal individual

abnormal chromosome numbers, recognition of the increased risk is based on empiric data (observations of at-risk populations). When the basic defect involves abnormal chromsome structure, the actual risk reflects either empiric data or, more frequently, principles of cell division and chromosome segregation. The designated risks for single-gene disorders are based on the principles of independent assortment of genes as defined by Gregor Mendel (3). The most important consideration is that the actual risk of recurrence depends on both the actual pattern of inheritance (autosomal recessive, autosomal dominant, or X-linked) and the relationship of the person concerned to the affected person. Thus, for a person with the autosomal dominant disorder of achondroplastic dwarfism, each of his or her offspring would have a 50% chance of being similarly affected, while a normal brother or sister's offspring would have only a slight risk, if any at all. For polygenic disorders, all first- and second-degree relatives would have increased risks compared to the general population. Occasionally it can be appreciated that a disorder is genetic, although specific genetic mechanisms may not be obvious—thus the need for the A4 subcategory (unspecified mechanism) in Table 2-2. For example, the cat-eye syndrome (4) is only sometimes associated with an extra chromosomal fragment in the patient's cells, and then it is not clear whether this disorder is chromosomal in nature (i.e., due to the extra chromosome) or whether the extra chromosome is merely an incidental finding. Another example is familial, nonspecific, moderate mental retardation. Based on the family pedigree and other information, it is not always possible to distinguish between single-gene and polygenic mechanisms.

The *possibly genetic* category (B) is very important because it represents those instances where, for one reason or another, the presence or absence of genetic factors cannot be appreciated and the likelihood of recurrence therefore cannot be clarified. In terms of providing practical assistance, there are really two sets of subcategories: those where a genetic component is probable (B1, B2), so that a degree of caution is warranted; and those where a genetic component is uncertain at best (B3, B4), so that essentially nothing can be said about the likelihood of recurrence. Disorders are included in the B1 "probable" subcategory because the *family history* suggests a genetic element that cannot be further characterized. Certain families with multiple instances of leukemia or of limb malformations (e.g., the Poland syndrome) fall into this group. In other situations the probability that genetic factors are involved is based on the *previous experience* of the clinician, or on a few, perhaps anecdotal or inconclusive, reports (B2). Many diagnoses are *descriptive* because a specific designation is not possible; they are placed in the *B3* "uncertain" subcategory. This category is important because it emphasizes, if not defines, the limits of our abilities—both in terms of diagnostic acumen and recurrence predictability. Many instances of multiple congenital anomalies or "learning disability" fall in this category. In some cases they are bona fide syndromes that have not been formally recognized and characterized; in others they are merely nonspecific combinations of features. If no previous cases with the same specific features are known, then placement in this subcategory is generally warranted. The implication is that with a specifiable cause, a pattern should be apparent; and conversely, with no pattern, there is no specified cause or recurrence risk. The importance of extensive clinical experience, consensus diagnosis, and sharing of experiences through conferences, journals, and computer registries becomes obvious: before a given diagnosis can be classified as "possibly genetic; uncertain for lack of specific diagnosis" (C3), review by one or more experts is mandatory. The final subcategory in the "possibly genetic" category is determined when a known or presumed *heterogeneity* of causes is responsible for the disorder (B4). The earlier example of a family's isolated case of aqueductal stenosis applies here. Instances of unilateral retinoblastoma might also be classified in this manner.

Nongenetic disorders can make up 20% of all the referrals to clinical genetics services (Table 2-3). Since the nongenetic nature of such disorders is usually not decided upon until the referral evaluation has been completed, the diagnostic part of the genetic counseling process is thereby emphasized. The vast majority of these disorders fall into the following general groups: single organ/tissue anomalies (e.g., phocomelia), syndromes due to exposure to potentially adverse environmental agents (e.g.,

TABLE 2-3
The Genetic Classification of 804 Patient Diagnoses Among CW-RGCP Referrals. The Number and Percent for Each Category and Subcategory Are Shown

Major Category	*No.*	*%*	*Subcategory*	*No.*	*%*
Genetic	528	66	A1. Chromosomal	105	13.0
			A2. Single gene	288	35.8
			A3. Polygenic	134	16.7
			A4. Unclear	1	0.1
Possibly genetic	112	14	B1. Probable/pedigree	14	1.7
			B2. Probable/experience	4	0.5
			B3. Uncertain/ nondiagnosis	78	9.7
			B4. Uncertain/ heterogeneity	16	2.0
Nongenetic	164	20	C1. Developmental	104	12.9
			C2. Environmental	19	2.4
			C3. Other, nongenetic	30	3.7
			C4. Normal	11	1.4

diphenylhydantoin), otherwise unexplained mental retardation or learning disability, certain neurodegenerative disorders (e.g., Gilles de la Tourette syndrome), hypogonadism, and recurrent abortions or infertility. *Developmental* disorders (C1) represent aberrations of embryonic development, manifested as single or multiple congenital malformations, explained neither by known or presumed genetic evidence, nor by known teratogens. *Environmental* disorders (C2) encompass two major clinical situations:

1. Clinical derangements retrospectively explained by in utero exposure to an environmental agent. Most often these involve congenital malformation syndromes, though occasionally the defect does not show itself until some time later [i.e., childhood-onset deafness secondary to congenital rubella (5), and the development of vaginal carcinoma in young girls exposed to diethylstilbestrol in utero (6)].
2. Situations where the concern about the environmental agent is prospective, i.e., before the birth, or even conception, of the at-risk fetus. Exposure to various amounts of diagnostic x-rays is a frequent source of concern.

If the referral problem cannot be explained by genetic or environmental factors and is not due to an actual developmental aberration, then it falls into the "other, nongenetic" subcategory (C3). These are essentially *acquired* disorders. An example would be a premature infant born the second of monozygous twins in the context of prolonged labor, respiratory distress, secondary disseminated intravascular coagulation, and resultant subarachnoid hemorrhage, which in turn leads to mental retardation and spastic quadriplegia. A modest number of referrals represent essentially *normal* children (C4) with one or two minor normal variations, such as simian creases. These "false-positive" referrals indicate the willingness of referring clinicians to implement the genetic approach to human disease through referral for the corroboration of a diagnosis.

Implementation of the Classification

Examples of the more common types of disorders found among patients referred to the Colorado-Wyoming Regional Genetic Counseling Program (CW-RGCP) during its first 29 months of operation are given in Tables 2-3, 2-4, 2-5, and 2-6. 772 families were referred to the program, but because there were occasionally two or more diagnoses per family, the total number of diagnoses was 804. These diagnoses are instructive because they do not merely reflect the activities of a tertiary-care genetics clinic in a large medical center. Rather, they reflect a patient population drawn directly from primary-care settings in suburban and rural

TABLE 2-4

General Types of Disorders Referred to the CW-RGCP for Genetic Evaluation and Counseling.[1] The Types of Disorders and Their Respective Percents Are Itemized in Descending Order of Frequency

1. Nonspecific mental retardation, failure to thrive and anomalies, singly or combined	17.2
2. Single organ, polygenic	15.5
3. Chromosomal, including amniocentesis	13.6
4. Neuromuscular	13.5
5. Skeletal	6.2
6. Biochemical, metabolic	5.3
7. Face, ear, nose, throat (no cleft lip/palate)	3.8
8. Obstetric, gynecologic	3.4
9. Ophthalmologic	3.1
10. Renal, genitourinary	2.9
11. "Syndromes"	2.8
12. Hematologic	2.8
13. Endocrinologic	2.4
14. Environmental	2.2
15. Dermatologic	1.8
16. General genetic (e.g., consanguinity)	1.7
17. Connective tissue	1.5
18. Gastrointestinal	0.3
	100.0%

[1]Note especially that the largest percentage (17.2%) is for the nonspecific descriptive diagnosis category and that polygenic disorders are the most common genetic problems (15.5%).

TABLE 2-5

Disorders Considered to Represent Nonspecific Mental Retardation, Failure to Thrive, and Anomalies, Singly or Combined—Broken Down According to Their Relative Proportion (%) and Genetic Classification Subcategories.

Type of Disorder	%	*Classification Subcategories*
Mental retardation, developmental delay, learning disability	39.2[1]	A2, A3, B1, B3, C1, C3
Multiple anomalies, developmental	28.2	C1
Multiple anomalies, diagnosis not otherwise known	17.0	B3
Failure to thrive, short stature	5.2	A2, B3, C1, C3
Stillborns, newborns—acquired disorder	5.2	C3
Children, adults—acquired disorder	4.4	C3
Dyslexia	0.8	A3

[1]This percentage emphasizes how often mental subnormality is a part of genetic and congenital disorders.

TABLE 2-6

More-or-less Specific Referral Diagnoses, Their Relative Contributions (per cent) to the CW-RGCP Patient Load (in Descending Order of Frequency), and Respective Genetic Classification Subcategories[1]

	%	
Down syndrome	5.9	A1
Cleft lip and/or palate	5.5	A2, A3
Neural tube defects	3.4	A3
Congenital heart disease	3.3	A3
Hearing deficit, deafness	2.5	A2, B1, B2, B3, C1
Chromosomal, prenatal diagnosis	2.4	A1
Neurofibromatosis	2.0	A2
Sex chromosome aneuploidy	1.7	A1
Seizures, major motor	1.7	A2, A3, B1, B2
Dwarfism	1.7	A2
Metabolic storage disease	1.7	A2
Chromosome breaks, rearrangements	1.5	A1
Limb anomalies, sporadic-development	1.4	C1
Limb reduction defects, other	1.4	A2, B2, B3
Cystic fibrosis	1.4	A2
Concern for fetal effects of drugs, irradiation, infection	1.4	C2
Duchenne muscular dystrophy	1.3	A2
Microcephaly	1.3	A2, A3, B1, C1
Brachydactyly, syndactyly, polysyndactyly	1.3	A2, A4, B2, B3, B4, C
Diabetes mellitus	1.3	A2, A3
Club foot	1.1	A3
Multiple miscarriages	1.1	A2, B1, B3
Urinary-tract anomalies	1.1	A2, A3, B1, B2, B4
Sickle-cell anemia	1.1	A2
Hydrocephalus	1.0	A3
Trisomy 13 (Patau syndrome)	1.0	A1
Huntington chorea	1.0	A2
Arteriosclerotic heart disease/hyperlipidemia	1.0	A2, A3
Polycystic dysplastic kidneys	1.0	A2, C1
Muscular dystrophy, other	0.9	A2
Progressive spinal muscular atrophies	0.8	A2
Charcot-Marie-Tooth disease	0.8	A2
X-linked mental retardation, neurodegeneration	0.8	A2
Retinal pigmentation disorders	0.8	A2
Hemophilia	0.8	A2
Epidermal, nonpigmentary disorders	0.8	A2
Minor variants in an otherwise normal child	0.8	C4
Tuberous sclerosis	0.6	A2
Infertility	0.6	B3, C4
Blood-group incompatibility	0.6	A2
Silver syndrome	0.6	C1
Viral teratogenesis	0.6	C2
Cutaneous pigmentary disorders	0.6	A2
Aminoacidopathies	0.6	A2
Trisomy 18 (Edwards syndrome)	0.5	A1
Branchial arch syndromes	0.5	A2, C1
Obstetric problems, acquired	0.5	C3
Strabismus	0.5	A2, B3
Cornelia De Lange syndrome	0.5	B4
Thyroid abnormalities	0.5	A2, B3, C4
Consanguinity, incest	0.5	A2
Noonan syndrome	0.4	A2

Thalassemia	0.4	A2
Nonspecific joint laxity syndromes	0.4	A2
Marfan syndrome	0.4	A2
Ehlers-Danlos syndrome	0.4	A2
Trisomy 8 (Warkany syndrome)	0.3	A1
Total	65.9%	

[1]This table gives an idea of the relative contributions of more-or-less specific diagnoses to our CW-RGCP clinic population, drawn from suburban and rural primary-care practices. Note that some categories may have several etiologies. Down syndrome was the most common specific basis for referral (not including amniocentesis). The second, third, and fourth categories are all polygenic traits, themselves adding to 12.2%.

communities. Table 2-4 provides the most general breakdown; diagnoses are grouped into broad types indicating etiology (e.g., chromosomal), major organ system involved (e.g., skeletal), or even broader, more descriptive diagnoses (e.g., "nonspecific mental retardation; failure to thrive; and anomalies, single or combined"). This type of organization is an attempt to make a large amount of data manageable for various purposes, and reflects actual referrals in a given time period (29 months). It is thus somewhat arbitrary in terms of both the inclusion of a given type of diagnosis and the limited number of types.

Nonspecific disorders are tabulated in Table 2-5. Their heterogeneity is indicated by the diversity of subcategories (A2, A3, B1, B3, C1, C3). They were present in more than half of the 182 patients in subcategories B3 and C1 in Table 2-3. The major point is that the diagnosis *per se* was inadequate for maximum refinement of recurrence-risk estimates. Further categorization did allow for refinement, however. Table 2-6 shows the relative contributions of specific diagnoses, or types of diagnosis, in the CW-RGCP referral population; it complements Table 2-3 by exemplifying disorders included in the major categories and subcategories. It also demonstrates how given disorders might be classified. Specific diagnoses, if known, were utilized only if their contribution to the whole was at least 0.4%. An exception was the Warkany syndrome (0.3%), which was included for special reasons to be discussed in Chapter 3. The less specific diagnoses reflect either the inability to be more specific (e.g., major motor seizures), the logical association of diagnoses based on similar organ-system involvement (e.g., congenital heart disease), similar pathogenetic considerations (e.g., amnioacidopathies), or similar clinical consequences (e.g., multiple miscarriages). This organization was adopted for the sake of brevity: to itemize each specific diagnosis would have been too unwieldy, and nothing would be added conceptually.

In closing this chapter I wish to reemphasize three important facts. First, 20% of our CW-RGCP referrals were for nongenetic disorders, and inclusion in this category was a consequence of the referral evaluation. It was the implementation of the genetic approach that led to clarification of the diagnosis and thus the low recurrence risk for these families. Second, in those cases where a genetic etiology could be specified, the basis for prevention was thereby established. With the consequent application of additional genetic methods—effective communication in genetic counseling, artificial insemination by donor, prenatal diagnosis, etc.—prevention can often become a reality (7). Third, although other geneticists might classify some of the diagnoses differently, such differences do not detract from the usefulness of the classification. If anything, they highlight the situations where expert consultation is needed.

Cited References

1. Day N, Homes LB: The incidence of genetic disease in a University hospital population. *Am J Hum Genet* 25:237, 1973.

2. Riccardi VM: Health care and disease prevention through genetic counseling: A regional approach. *Am J Publ Hlth* 66:268, 1976.

3. Stern C: *Principles of Human Genetics* (3rd ed.). San Francisco, WH Freeman, 1973, pp 145–186.

4. Cory CC, Jamison DL: The cat eye syndrome. *Arch Ophthal* 92:259, 1974.

5. Menser MA, Doos L, Harley JD: A twenty-five-year follow-up of congenital rubella. *Lancet* 2:1347, 1967.

6. Herbst AL, Poskanzer DC, Robboy SJ, et al: Prenatal exposure to stilbesterol. *N Engl J Med* 292:334, 1975.

7. Milunsky A, Atkins L: Prenatal diagnosis of genetic disorders. in Milunsky A (ed): *Prevention of Genetic Disease and Mental Retardation.* Philadelphia, W B Saunders, 1975, pp 221–263.

General References

A. Martin GM, Hoehn H: Genetics and human disease. *Hum Path* 5:387, 1974.

B. McKusick VA: The nosology of genetic disease. in McKusick VA, Claiborne R (eds): *Medical Genetics*, New York, HP Publishing Co., 1973, pp 211–220.

CHAPTER THREE

GENETIC MECHANISMS OF DISEASE—CHROMOSOME DISORDERS

General Considerations

Genetic disorders, by definition, are due to aberrations of the genetic material. The genetic material itself has various levels of organization: 1. chemical—arrangement and sequences of molecules; 2. functional—genes, which are expressed as specific traits or physiological activities; and 3. chromosomal—aggregations of genes into microscopically identifiable units. Although any given genetic disorder can ultimately be discussed in terms of the first level of organization—chemical—each is usually first recognized as being genetic through changes in gene function and expression or in chromosome structure. A change in the functional expression of a gene is a mutation. Mutant genes are expressed as deviations from the normal in such a way that recurrence among certain future family members is predictable. In addition, the immediate gene product, or its deficiency, is often detectable, allowing for specific diagnosis and the possibility of screening, detection of the carrier state, and prenatal diagnosis. Aberrations in the larger organizational units—chromosomes—involve structural rearrangements or changes in chromosome numbers.

Chromosome Disorders—An Introduction

Chromosome disorders are dealt with first for many reasons: 1. They are by definition genetic disorders. 2. They are common, being the most frequent specified cause for genetic referrals. 3. The potential for prevention through amniocentesis is readily appreciated. 4. There have been many new technical advances in delineating chromosome structure, function, and ability to produce disease. 5. The principles for ascribing recurrence risks, both Mendelian and empiric, apply to other genetic disorders as well.

Within fairly narrow limits, too much or too little chromosomal material in all or some of an individual's cells leads to clinical disorders. Most, if not all, of the congenital syndromes associated with chromosomal aberrations are considered to be caused by the chromosome abnormality; the same is true of multiple miscarriages and infertility in certain cases. In other situations, exemplified by the well known association of the Philadelphia chromosome with chronic myelogenous leukemia, it is not yet clear whether the apparently acquired chromosome changes are incidental to the disorder or a cause of it. Dozens, if not hundreds, of chromosome disorders have been described (1), and it is becoming evident that many of these disorders may have more than one chromosomal explanation (2). Moreover, the availability of amniocentesis for the prenatal diagnosis of chromosome disorders has added a compelling reason for establishing a correct chromosomal diagnosis. Thus, with broader indications for chromsome analysis and more salutory consequences of clarifying the chromosomal aberration, clinicians in all disciplines need to be familiar with the various principles involved. This section will clarify those principles, characterize the importance of chromosome analysis on a clinical level, and develop guidelines for applying the principles. With some exceptions the clinical features of the disorders will not be discussed; for those the reader is referred elsewhere (3,4).

In the most general terms there are four basic reasons for obtaining a chromosome analysis on behalf of a given patient:

1. To delineate the chromosomal mechanism that accounts for a clinical diagnosis even when the etiology is known to be chromosomal. For example, it is imperative to know whether the underlying mechanism for the Down syndrome is an extra intact chromosome 21 (aneuploidy) or a chromosomal rearrangement (translocation) which may have been inherited from a parent.

2. To establish the etiology and thereby make a specific diagnosis in the face of an inadequate descriptive diagnosis. Increasing numbers of mental retardation—multiple malformation disorders are being shown to result from chromosome rearrangements of a subtle and/or unusual nature. The disorders described below involving loss of material from chromosome number 11 (5) and cases where either infertility or multiple miscarriages are shown to be due to a chromosome abnormality would exemplify such rearrangements.

3. To establish another parameter of a disorder when a specific diagnosis has already been made. This objective applies where there are two or more concurrent diagnoses (6) or where a chromosome aberration would serve as a marker with prognostic significance, i.e., the Philadelphia chromosome in chronic myelogenous leukemia or an extra chromosome 8 in acute myelogenous leukemia (7).

4. For purposes of clinical investigation, usually as an attempt to localize specific genes to specific chromosomal segments.

Principles, Terminology, and Techniques

Before dealing more specifically with chromosome disorders and how chromosome analysis fits into health care efforts, it is worthwhile to review certain terms and concepts (see also ref. 8). A glossary of common terms is provided in Table 3-1. The term *chromosome disorder* will refer to the clinical manifestations of the variation from normal in a person's chromosome constitution. The terms *chromosome abnormality, chromosome anomaly, chromosome defect,* etc. will refer to the actual change in chromosome number and/or structure. A chromosome disorder is caused by a chromosome abnormality. Thus, for a chromosome disorder,

TABLE 3-1
A Brief Glossary of Chromosomal Disorders

Acquired: Noncongenital and nonheritable.
Acrocentric: Chromosome with the centromere very close to one end, making the short arm very small; normally includes chromosome numbers 13, 14, 15, 21, 22, and Y.
Aneuploidy: Abnormal chromosome number; applies to humans when the chromosome number in somatic cells is other than 46.
Autosome: Non-sex chromosome; any member of chromosome pairs 1 through 22.
Band: Specific markings along the length of a chromosome after chromosome preparations are treated and stained in special ways; The alternating dark and light bands give characteristic patterns for each chromosome and segment thereof.
Banding techniques: Various methods of manipulating chromosome preparations to elicit specific banding patterns along the length of each chromosome; especially G, Q, R, and C banding.
Barr body: Same as X-chromatin mass.
Centromere: That segment of the chromosome where sister chromatids are attached to each other; that section of the chromosome or chromatid which separates the short arm from the long arm; the site of spindle attachment in meiosis and mitosis.
Chromatid: The longitudinal half of a chromosome as it appears in metaphase; sister chromatids are the complementary halves of a given chromosome, each of which separates from the other by the end of cell division; after separation each is properly referred to as a chromosome itself.
Chromosome: Microscopically identifiable units of genetic material (chromatin); each species has a characteristic number and pattern of chromosomes; the functional units of genetic material—genes—are linerally arranged along the length of each chromosome.
Congenital: Present at the time of birth; may or may not be genetic.
Crossing over: The process by which nonsister chromatids of homologous chromosomes undergo reciprocal exchange during the first division process of meiosis.
Deletion, interstitial: Loss of a chromosomal segment other than the terminal portions of the chromosome; implies at least two chromosomal breaks.
Deletion, terminal: Loss of a terminal segment of a chromosome; implies a single chromosomal break or a translocation.
Diploid: Twice the number of haploid chromosomes ($2n$), including two members of each pair; the number of chromosomes normally present in a somatic cell nucleus, 46 in man.

TABLE 3-1 (*Continued*)
A Brief Glossary of Chromosomal Disorders

Distal: In chromosome terminology, that part or parts of the chromosome relatively farther from the centromere and closer to the ends (opposite of proximal).
Euchromatin: The genetically active portion of chromatin.
Inversion: A chromosomal rearrangement wherein a given segment of chromosome rotates 180° about the fragment's midpoint so that its distal end is now proximal and vice versa, except for inversions about the centromere wherein the short arm portion adjoins the long arm and vice versa.
Haploid: The number of chromosomes found in a normal germ cell, representing one of each pair; designated by the letter *n;* 23 in man.
Heterochromatin: The genetically inactive portion of chromatin.
Karyotype: The ordered arrangement of a cell's chromosomes according to established conventions, currently those stipulated by the Paris Conference.
Long Arm: The longer of the two normal chromosome arms when they are disparate; currently this definition is supplemented by the specifications of the banding patterns according to the karyotype idiogram from the Paris Conference; the long arm is designated by the lower case letter q.
Lyon principle: When a cell has more than one X-chromosome, all but one of them become functionally inactive and stain darkly in interphase (Barr body). The initial inactivation occurs early in embryonic development and is random with respect to the parental origin of the X-chromosome involved. Once inactivated, an X-chromosome and all those derived from it in cell division remain so.
Meiosis: The process of cell division leading to mature germ cells that ultimately have the haploid number of chromosomes.
Metacentric: Chromosomes having short and long arms of essentially equal lengths, so that the centromere is just about in the middle of the chromosome.
Metaphase: That part of the cell cycle in which the chromosomes are most suitable for routine analysis; each chromosome is condensed and all are situated in a single plane (plate); between the prophase and anaphase segments of the cell cycle.
Mitosis: The process of cell division involved in duplication of somatic cells, whereby the diploid number of chromosomes is preserved in all daughter cells.
Monosomy: The presence of only one sex chromosome or only one member of a pair of autosomes in a somatic cell.
Mosaicism: The presence, in a given individual, of two or more cell lines, each characterized by a distinctive karyotype.
Nondisjunction: The failure of sister chromatids to disjoin or go separate ways in the anaphase of mitosis or the second stage of meiosis; or the failure of homologous chromosomes to disjoin in the first stage (reduction division) of meiosis.
Paris Conference: At the 1971 International Conference of Human Genetics in Paris, the current system of chromosome terminology and conventions was adopted, based on early experience with chromosome banding techniques.
Partial monosomy: The presence of a chromosomal segment on only one homolog, the corresponding segment on the other homolog having been deleted or translocated.
Partial trisomy: The presence of a chromosomal segment in triplicate.
Polyploidy: An excess of chromosomes per cell so that the total number is a simple multiple of the haploid number (e.g., triploidy = $3n$, tetraploidy = $4n$).
Polysomy: The presence of a chromosome in a cell more than the normal number of times, e.g., trisomy 21.
Proband: The affected individual whose disorder prompts family studies and genetic counseling.
Proximal: In chromosome terminology, the part or parts of the chromosome relatively closer to the centromere and farther from the ends (opposite of distal).
Satellites: Chromosomal material representing the distal segments of the short arm of acrocentric chromsomes; their appearance with most cytologic stains falsely suggests that they are not directly attached to the short arm. Distinctly staining satellites can serve as useful chromosome markers.
Sex chromosomes: The X and Y chromosomes.

Short arm: The shorter of the two chromosome arms when they are disparate; currently this definition is supplemented by the specifications of the banding patterns according to the karyotype idiogram from the Paris Conference; the short arm is designated by the lower case letter p.
Submetcentric: Chromosomes having short arms distinctly shorter than the long arms, so that the centromere is displaced from the middle of the chromosome.
Translocation, reciprocal: The mutual exchange of chromosomal segments between any two chromosomes. In the person where this has occurred, there is no net loss of chromosomal material, but the production of chromosomally unbalanced gametes becomes possible.
Translocation, Robertsonian: The mutual exchange of chromosomal material in a special way between two acrocentric chromosomes (numbers 13, 14, 15, 21, 22). The net result is that the two chromosomes' long arms are attached to each other by a centromere, and all or most of their short arms and one centromere have been lost. There is no adverse effect of the short-arm loss, but, in contrast to a reciprocal translocation, there is one less centromere, and therefore one less chromosomal character in the karyotype.
Translocation carrier: A person whose somatic cells demonstrate either translocation rearrangement noted above. This person may be sterile, but if not, is then capable of transmitting an unbalanced portion of the rearrangement.
Trisomy: The presence of three chromosomes instead of a pair, e.g., trisomy 21.
Sex chromatin: Same as X-chromatin.
X-Chromatin: An inactive X-chromosome (see Lyon principle) is represented by a darkly staining nuclear mass throughout most of the cell cycle (interphase). The number of X-chromosomes in a cell is equal to the number of X-chromatin masses plus one.
Y-Chromatin: A Y-chromosome fluoresces with a distinct brilliance in interphase of the cell cycle when quinacrine is used as the cell stain. This bright mass is the Y-chromatin body. The total number of Y-chromatin bodies represents the number of Y-chromosomes.

there are always two levels of diagnosis. For example, the clinical diagnosis of the Down syndrome (chromosome disorder) may result from any one of several chromosome abnormalities which must be demonstrated in its own right (chromosome diagnosis).

A normal cell's chromosomes, fixed and stained in metaphase of the cell cycle, are shown in Fig. 3-1. The magnification is about 6500X. Each chromosome is then cut from the photograph and arranged according to standards set by the 1971 Paris Chromosome Conference (9,10). In its simplified form the symbolic designation of chromosome constitution (karyotype) is easily understood. The total number of separate chromosomes per cell is given by an arabic number (one chromosome for each centromere). If relevant, the sex chromosome constitution is indicated by one X and/or Y for each respective sex chromsome. A normal male karyotype is thus 46,XY (Fig. 3-2) and a normal female 46,XX (Fig. 3-3). If the sex chromosomes are normal these can be left out and changes in autosomes noted directly. If a chromosome is extra, its number is preceded by a plus (+) sign; if missing, it is preceded by a minus (−) sign. Trisomy 21 is thus designated as 47,+21 (Fig. 3-4). If there is extra material on a chromosome arm, it is designated as,

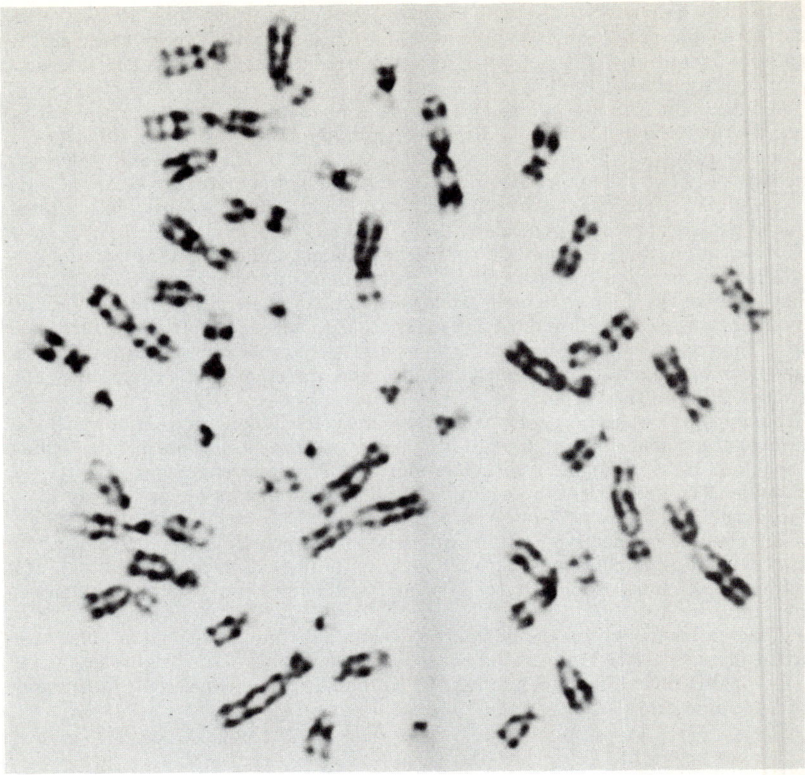

Figure 3–1 G-banded 46,XX metaphase spread, at about 6500X magnification. For further analysis, each chromosome is individually cut out of the original photographic print and then arranged in a karyotype.

for example, 14q+, 7p+ (respectively, extra material on the long arm of #14 and short arm of #7). If material has been deleted it is noted by a minus sign following the number and arm sign, i.e., 4p−, 11p− (respectively, deletions of the short arm of #4 and short arm of #11). The signatures for other structural rearrangements can be more complicated. For example, a translocation between the short arm of #7 and the short arm of #20 is denoted by a "t" followed by parentheses containing the symbols 7p and 20p (lowest number first), separated by a semicolon: t(7p;20p). Several chromosomes have, as a normal variation, a decreased or increased amount of densely staining chromatin near the centromere of the long arm (qh region); this is designated respectively as qh− and qh+. Mosaicism is designated by the symbol for the karyotype of the cell line with fewest chromosomes followed by a slash (/), then the karyotype of the next largest cell line, and so on. For example, the designation for mosaicism involving X-monosomy and a normal sex chromosome complement is 45,X/46,XX. The designation for mosaic trisomy 8 (male patient) is 46,XY/47,XY,+8. Beyond this point these symbols become more complex and need not be pursued here (9,10).

Chromosome abnormalities are of two basic types: changes in chromosome number (more or less than 46 chromosomes per somatic cell) and changes in the structure of one or more chromosomes. *Monosomy* for an autosome is found in spontaneous abortions; its occurrence without mosaicism in a newborn infant has been suggested only for chromosome #21 (11). Monosomy for an X-chromosome occurs in both spontaneous abortions and live-born infants. The resulting disorder is widely variable; the clinical expression is determined in large part by the presence or absence of mosaicism for a normal (46,XX or 46,XY) cell line and/or another aneuploid cell line (e.g., 47,XXX, 47,XXY). *Polysomy* is known for sex chromosomes and several autosomes, especially

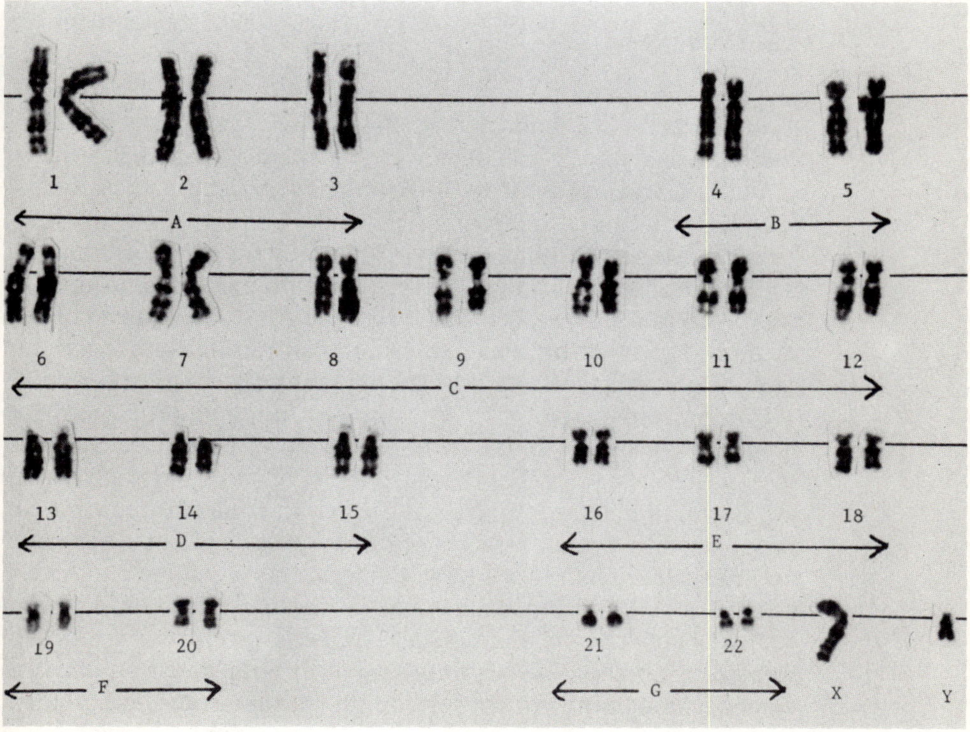

Figure 3–2 Normal male karyotype—46,XY.

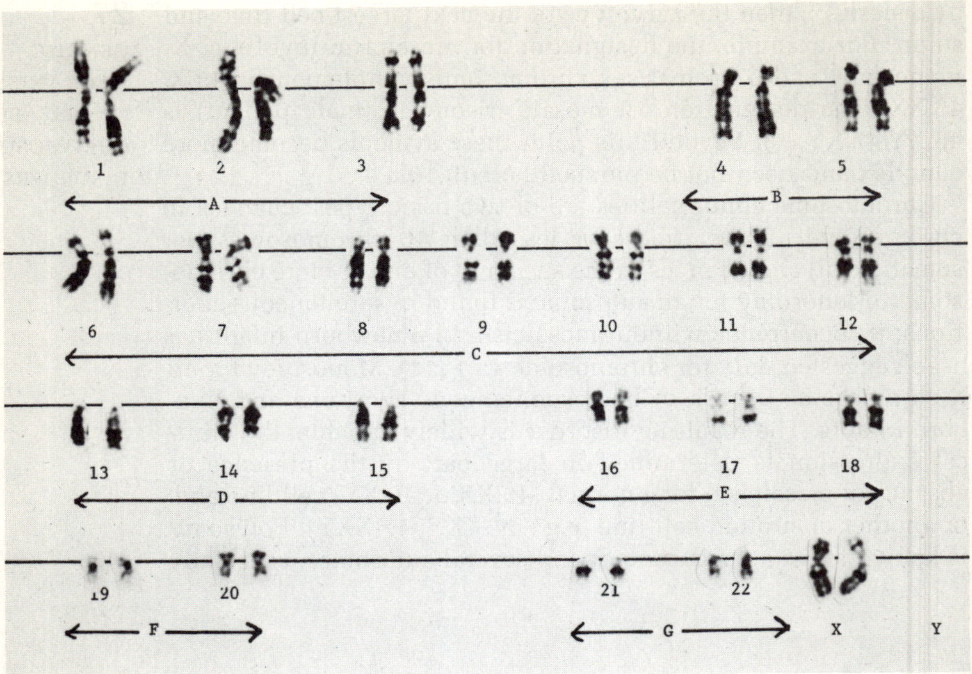

Figure 3–3 Normal female karyotype—46,XX.

numbers 8, 13, 18, and 21 (Fig. 3-4). When there are extra autosomes, it is always a trisomy. Mosaicism for a normal cell line is virtually always present with trisomy 8 and occurs occasionally with the other three commoner trisomies. Trisomy for a portion of a chromosome (partial trisomy) can occur by translocation (see below) or by the additional presence of a deleted chromosome, as for example 47,XX,+14q− (12). *Polyploidy*, the presence of one or more entire additional haploid sets of chromosomes, is very rare: it is mainly restricted to spontaneous abortions (13).

Chromosome rearrangements are best understood in terms of breakage. A simple break in one chromosome with subsequent loss of the distal segment (because it has no attached centromere) is a *terminal deletion*. Two breaks involving one chromosome can produce an *inversion*, loss of the intervening segment resulting in an *intestitial deletion* or loss of the terminal segments and formation of a *ring chromosome*. Breaks in two chromosomes and mutual exchange of their distal segments produce a *reciprocal translocation* (Fig. 3-5). If only one of the translocation chromosomes is transmitted at conception, the chromosome complement is then unbalanced in terms of deletion and/or partial trisomy. A special type of translocation is referred to as a *Robertsonian* or *centric-fusion translocation* (Fig. 3-6). It occurs only with acrocentric chromosomes and entails loss of one of the translocation products. The long arms of two acrocentrics become fused at a common centromere; so do their short arms but these are apparently unimportant genetically and are lost from the cell. *Isochromosomes* result from duplication of one arm of a chromosome and loss of the other arm. Thus, an isochromosome for the long arm of the X consists of two long arms joined at the centromere, with no short arm present. This configuration could arise by means of an X;X translocation, but it is usually thought to result from transverse splitting of the centromere. *Chromosome breaks* without

(a)

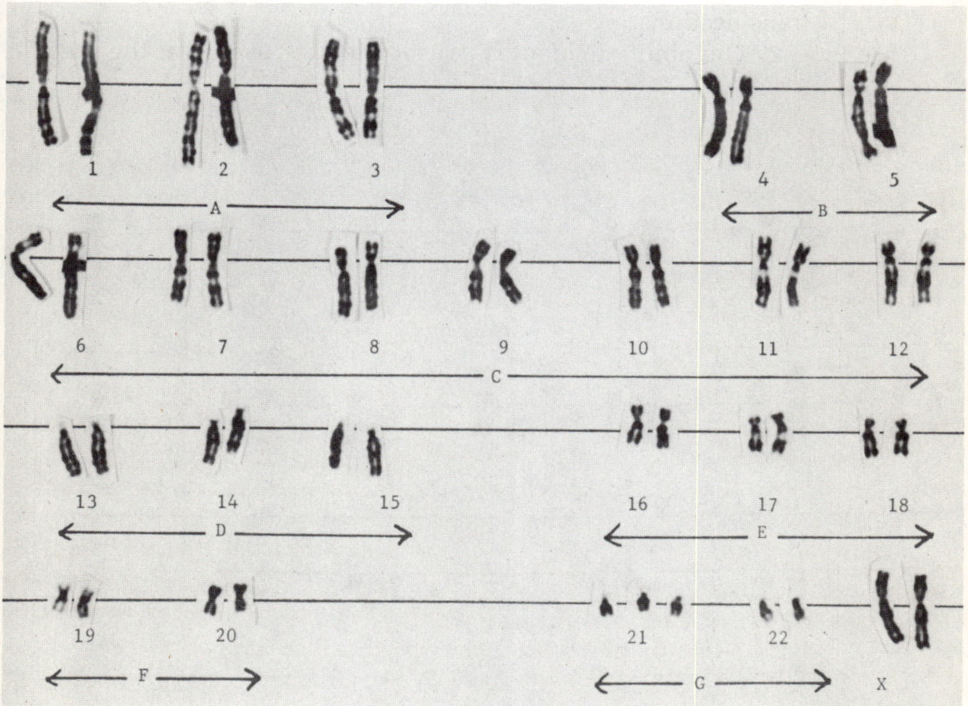

(b)

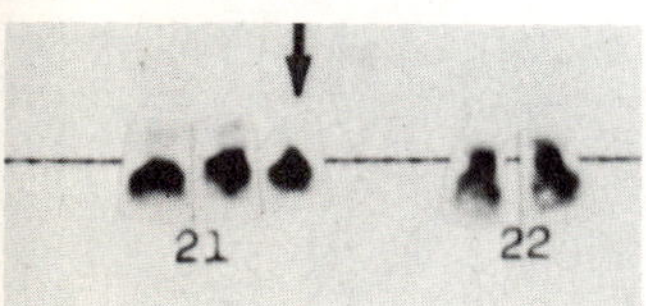

Figure 3–4 Trisomy 21. (a) Complete G-banded karyotype showing 47,XX,+21; (b) partial karyotype emphasizing the banding differences between chromosomes 21 and 22.

rearrangements (Fig. 3-7) entail a distinct separation between the broken elements and displacement of them; they may involve one or both chromatids. *Chromatid gaps* appear as small separations (nonstaining segments) between portions of single chromatids. *Sister chromatid exchanges* are normal occurrences, i.e., mutual exchanges of corresponding parts of sister chromatids of a chromosome. They are demonstrated by specialized techniques and may occur more frequently in certain heritable disorders (14).

An understanding of these mechanisms has become increasingly important:

1. As the indications for chromosome analysis have broadened, we are detecting more deletions, partial trisomies, and reciprocal translocations.
2. Our ability to identify rearrangements, as well as the specific

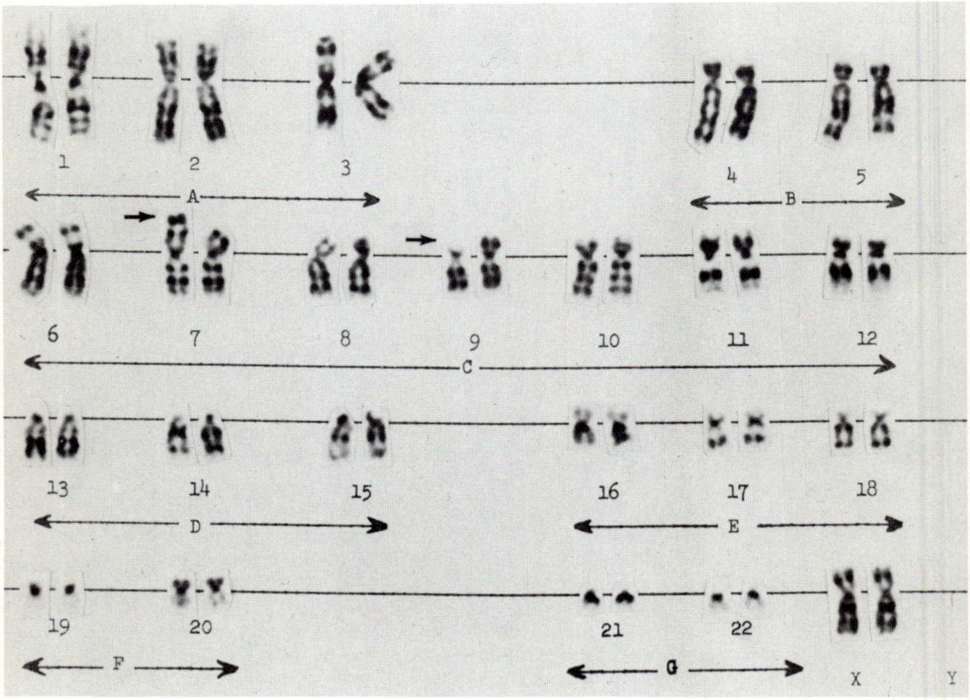

(a)

Figure 3–5 Reciprocal translocation. (a) Balanced. Note the extra material on the distal short arm of a #7 and loss of material from the distal short arm of a #9 (arrows); 46,XX,t(7p;9p).

segment involved in rearrangements, has vastly improved as a result of the banding techniques used in chromosome analysis (Table 3-2).

3. Once a balanced translocation carrier is identified, other potential carriers in the family should be studied to minimize the likelihood of bearing more children with the unbalanced translocation. Balanced carriers are usually recognized because an unbalanced offspring has been identified or an evaluation performed for multiple miscarriages or apparent infertility. Occasionally it is discovered incidentally when chromosomes are analyzed for purposes such as a random survey. The possibility of preventing recurrence among offspring to cousins, nephews, nieces, and siblings of the proband is a cogent reason for doing a chromosome analysis on a child with unexplained mental retardation, especially if it is associated with multisystem anomalies.

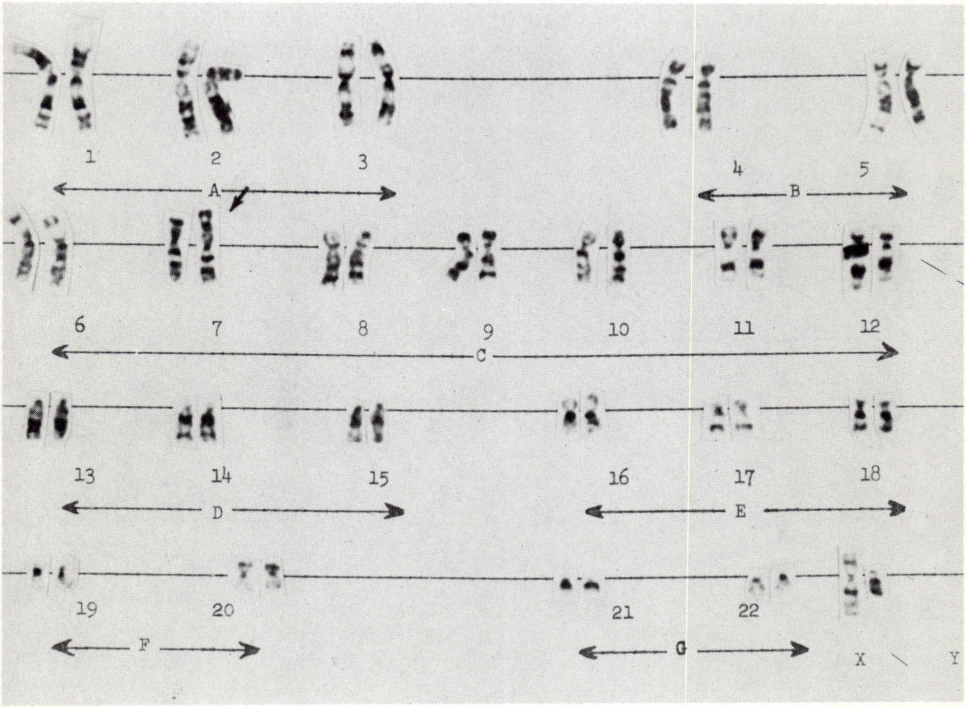

(b)

Figure 3–5
(b) Unbalanced. The son of the woman in (a), with the translocated #7 (arrow) and two normal #9 chromosomes, making the child effectively trisomic for the distal short arm of #9: 46,XY,+9p. (Courtesy of D. Peakman and A. Robinson.)

Such a possibility is not just hypothetical; it is a basis for preventive medicine being implemented with increasing frequency.

The banding techniques referred to above have helped to make chromosome analysis much more sensitive and thus a more meaningful part of day-to-day clinical practice. Cytogenetic laboratories around the world have reported (15) the detection of chromosome abnormalities in patients thought to be chromosomally normal by the old techniques. Before we deal with the banding techniques *per se*, however, it should be made clear what tissue can be used for chromosome analysis and how specimens are collected and processed. As indicated in Table 3-3, chromosome constitution can be assessed either directly or indirectly. The two approaches should be used to complement each other, not as mutually exclusive alternatives. The indirect methods currently apply only to sex chromosomes. They are most useful as screening devices and in establishing the presence and degree of mosaicism for sex chromosome aneuploidy. Moreover, both the X-chromatin and Y-chromatin status should be characterized. If

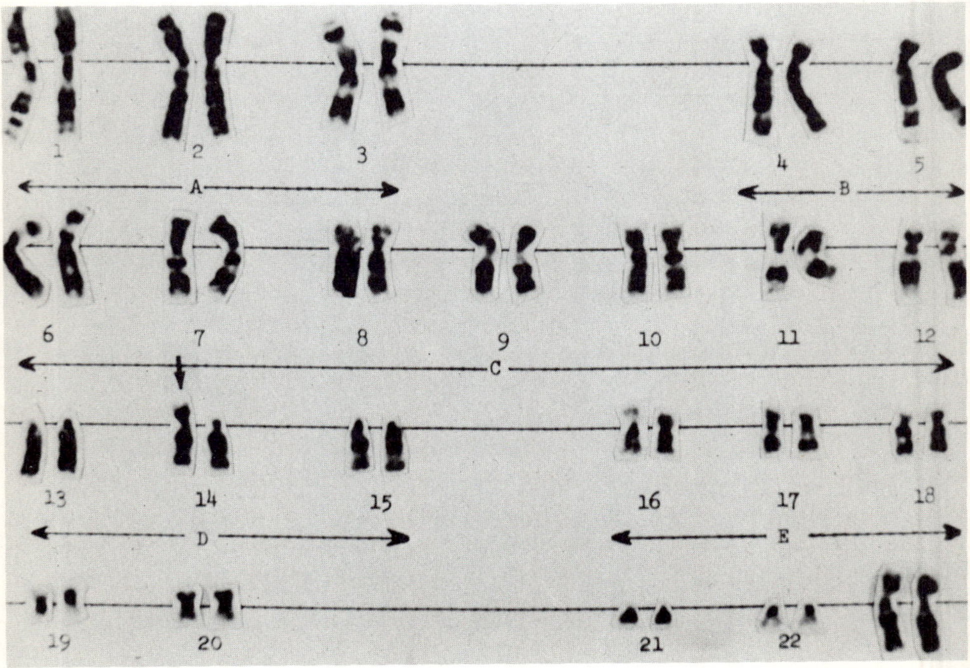

Figure 3–6 Robertsonian (14;21) translocation; unbalanced, from a patient with the Down syndrome. This girl with Down syndrome has extra chromosome 21 material attached to a #14. 46,XX,t(14q;21q).

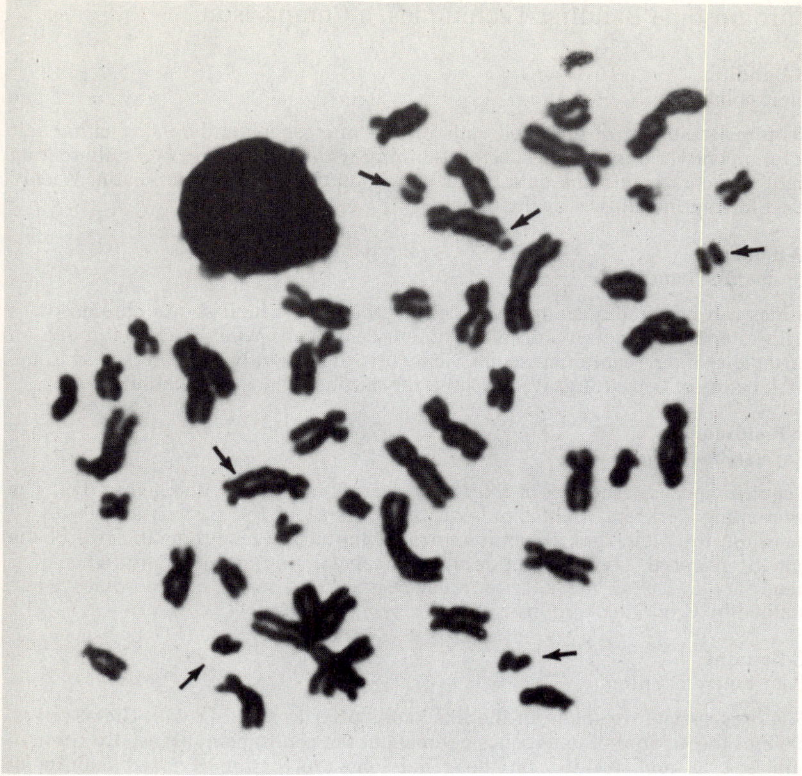

Figure 3–7 Chromosomal breakage. The arrows demonstrate chromosome fragments and a chromatid gap and chromosome break (top center). (Courtesy of D. Peakman and A. Robinson.)

the sex chromosome constitution is in question, then the answer relates to both sex chromosomes, not just one. Unfortunately, however, relatively few service laboratories perform Y-chromatin analysis. The number of X-chromosomes in a cell can be determined by nuclear staining: the inactive X-chromosome (vis-à-vis the Lyon principle) stains as a dense mass along the inside of the nuclear membrane (Fig. 3-8). The total number of X-chromosomes is equal to the number of X-chromatin masses (Barr bodies) plus one. Thus, for example, a normal female will have one Barr body and two X-chromosomes. The number of Y-chromosomes is equal to the number of brightly fluorescent nuclear masses seen when cells are stained with quinacrine and viewed through an ultraviolet fluorescence microscope. A normal male thus has one fluorescent Y-body (Fig. 3-9). These techniques are usually applied

TABLE 3-2

Chromosome Banding Techniques, a Comparison

G-Banding
Giemsa banding:

Requires treatment of the fixed cells on the microscope slide, using either salt solutions or various enzymes, before staining with Giemsa dye. Gives alternating light and dark bands, the dark bands corresponding to heterochromatin. Widely used for routine clinical evaluations.

Q-Banding
Quinacrine banding:

Slides can be directly stained with quinacrine dye, which shows chromosomes with alternating brightly and dimly fluorescent bands when viewed through an ultraviolet fluorescent microscope. Generally, the bright bands correspond to the dark bands of G-banding. Widely used for routine clinical evaluations.

R-Banding
Reverse banding:

Requires special treatment of fixed cells prior to staining with Giemsa dye. The alternating dark and light bands are the reverse of the pattern seen with G-banding; thus, the dark R-bands represent the active euchromatin form of the genetic material. Useful as an adjunct in special studies, especially when it is necessary to define the pale terminal segments of many chromosomes when stained by G or Q techniques.

C-Banding
Centromere banding:

Requires special treatment of the fixed cells prior to staining with Giemsa dyes. Defines the amount of heterochromatin about the centromere (especially chromosomes 1, 9, and 16) and on the distal long arm of the Y. Used mainly for investigational purposes, but since variations in these chromosome segments are very common, this technique has become increasingly useful in a number of ordinary clinical situations.

TABLE 3-3

Clinically Useful Techniques for the Direct and Indirect Assessment of Chromosome Constitution

1. Direct:
Karyotyping—actual depiction of chromosome numbers and structure by culturing cells, especially peripheral lymphocytes.

2. Indirect:
   a. X-chromatin (Barr body) determination—cell-nucleus stains of fixed, nondividing cells indicate the number of X-chromosomes. (The number of X's is the number of Barr bodies plus one.)
   b. Y-chromatin determination—fluorescing quinacrine stains of fixed nondividing cells indicate the number of Y-chromosomes. (For each brightly fluorescent spot there is one Y-chromosome.)

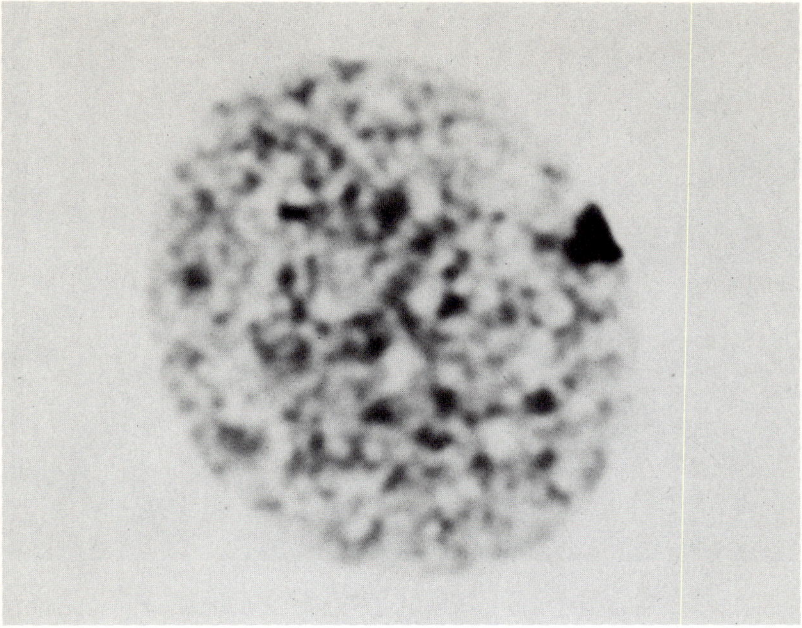

Figure 3–8 X-chromatin mass (Barr body). The large, darkly staining triangular mass along the nuclear membrane is a Barr body. It indicates that this cell has two X-chromosomes. (Courtesy of D. Peakman and A. Robinson.)

to scrapings of the buccal mucosa ("buccal smear"), although they are also useful for vaginal smears, amniotic membranes, and cultured skin fibroblasts. The ease of preparation and the large number of cells which can be examined makes these indirect methods suitable for screening and assessment of mosaicism. However, the preparations must be processed in an established cytogenetics laboratory that uses these procedures frequently; in less experienced laboratories there are too many false-positive and false-negative results.

Direct chromosome analysis can be done on any tissues whose cells will divide in culture (Table 3-4). The tissue most commonly used is peripheral venous blood, the T-lymphocytes of which can be made to divide by adding phytohemagglutinin to the culture medium (Fig. 3-10). Once the cells have been fixed and spread on slides, they must then be stained directly with quinacrine or pretreated before using other stains to elicit banding patterns on the chromosomes (Table 3-2). The two most widely used banding techniques are Q- and G-banding (16,17,18). With these techniques, it is possible to specifically identify each chromosome and its parts. These techniques have significantly increased the sensitivity of chromosome analysis, and have also made the analysis

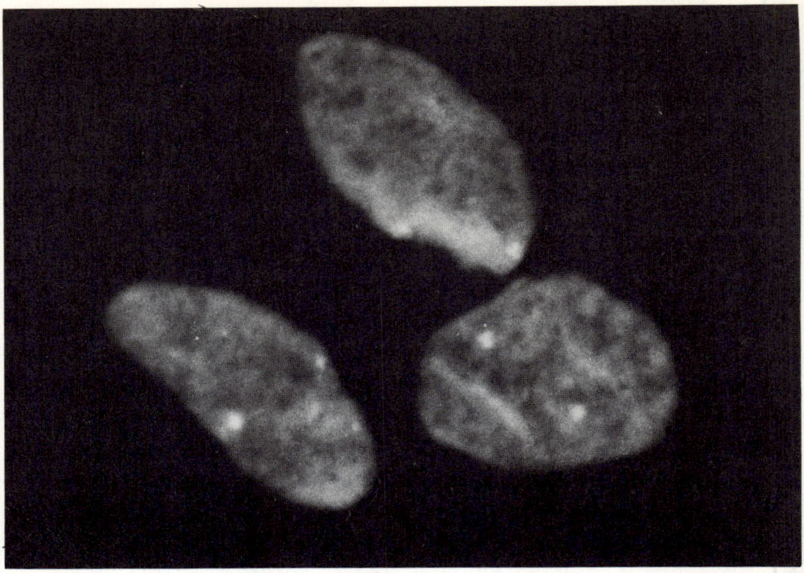

Figure 3–9 Y-chromatin mass. Three cell nuclei, each showing two brightly fluorescent spots, indicating two Y-chromosomes per cell. (Courtesy of D. Peakman and A. Robinson.)

more specific, allowing for more precise diagnosis of syndromes. After the syndrome characterized by Warkany in 1962 (19) was shown to be due to an extra chromosome 8, for example, the diagnosis was applied to at least 70 patients in a short period of time (20). As the recognition and characterization of chromosome abnormalities continue to improve, with the help of chromosome registry programs (21), the families of an affected patient can be

TABLE 3-4
Tissues Useful for Chromosome Preparations (Karyotyping)[1]

	Premortem	*Postmortem*
Peripheral lymphocytes	+	–
Bone marrow	+	–
Thymus	–	+
Spleen	–	+
Skin	+	+
Gonads	+	+
Amniotic fluid	+	–

[1]For postmortem sampling of the thymus and spleen for chromosome analysis, the specimens must be obtained before refrigeration. Skin specimens may be obtained up to 72 hours after refrigeration. A blood sample obtained at the moment of death may be processed as a premortem specimen.

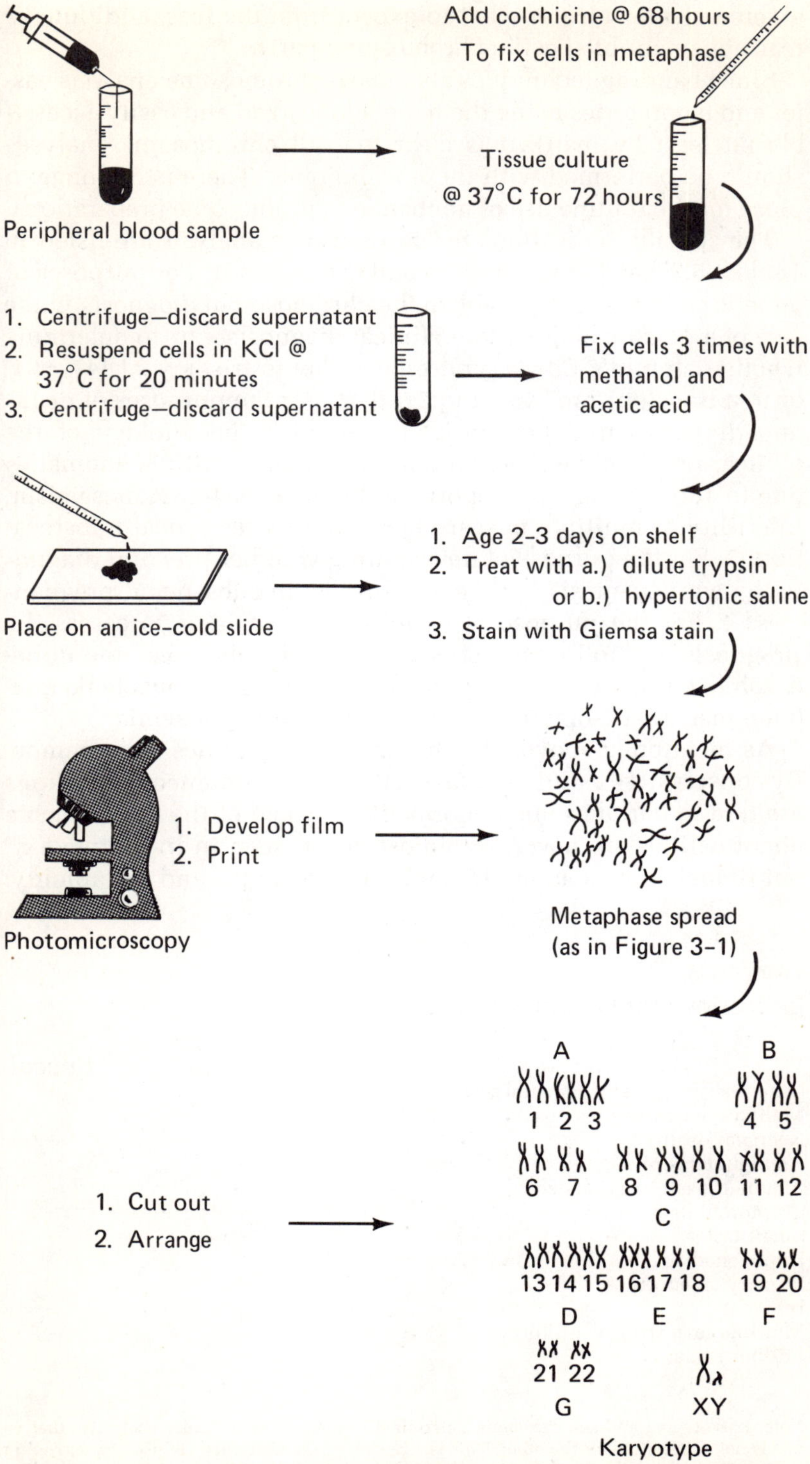

Figure 3–10 The processing of peripheral blood specimens for chromosome analysis.

given a clearer idea of what to expect from the loss, addition, or rearrangement of specific chromosome parts.

Since banding techniques also make chromosome analysis easier and laboratories using them are widespread and easily accessible (at least by mail), it is clear that all chromosome analyses should be performed with these techniques. There is no longer a place for the routine use of nonbanded chromosome preparations.

The specific indications for chromosome analysis are listed in Table 3-5. They fall into four eneral categories: 1. For purposes of *genetic counseling*, to establish the *chromosomal diagnosis* in the face of a presumed or certain clinical diagnosis, i.e., to determine whether a child's Down syndrome is due to trisomy 21 (47,+21) or to a translocation [46,t(14q;21q)]. 2. For the purposes of *accurate diagnosis and prognosis*, to establish the *etiology* of the clinical problem, i.e., mental retardation and multiple anomalies due to 18q−, primary amenorrhea due to 45,X/46,XX mosaicism, infertility or multiple miscarriages due to a reciprocal translocation. 3. For the purpose of determining whether *prenatal diagnosis* by amniocentesis will be appropriate in subsequent pregnancies. 4. For the purpose of establishing a *marker* to assess the *prognosis* and follow the *course* of a malignancy, i.e., the Philadelphia chromosome (Ph^1) in chronic myelogenous leukemia, bone marrow trisomy 8 in acute myelogenous leukemia.

As mentioned above, chromosome abnormalities are common. By conservative estimate at least 20% of all spontaneous abortions are due to chromosome abnormalities. Some of the chromosome abnormalities are never, or almost never, seen in liveborns, e.g., autosomal monosomies, trisomy 16, triploidy, and tetraploidy.

TABLE 3-5
Indications for Chromosome Assessment

	Karyotype	*Buccal*
Otherwise unexplained mental retardation	+	−
Multisystem anomalies	+	−
Neonatal death	+	−
Ambiguous genitalia	+	+
Inguinal masses in females	−	+
Cryptorchidism	−	+
Small testes	+	+
Hypogonadism	+	+
Primary amenorrhea	+	+
Infertility	+	−
Multiple miscarriage, stillbirths	+	−
Stillborn infant	+	−

Note: Karyotyping and sex chromatin determinations are never mutually exclusive, that is, one is not a substitute for the other. This is especially true when they are used for screening (e.g., ambiguous genitalia, primary amenorrhea) or when sex chromosome mosaicism is possible.

Others are seen with varying frequency in liveborn infants—45,X, trisomy 8, trisomy 13, trisomy 18, trisomy 21, sex chromosome polysomy, and a variety of deletions and unbalanced, as well as balanced, translocations. Among liveborn infants the frequency of chromosome abnormalities of all types is about 5 per 1000 (0.5%). In a number of high-risk groups, however, the contribution of congenital chromosome abnormalities is considerably higher (24) (neonatal deaths, multiple anomalies with mental retardation, multiple miscarriages, infertility without obstetric reasons or with sperm abnormalities, maternal age of 35 years or more, etc.). The frequency of acquired chromosome abnormalities is not known, and when they are found their significance is not clear. They are usually associated with certain viral infections, such as rubella; relatively large amounts of ionizing radiation exposure; certain malignancies, especially of the blood-forming elements; or with certain inherited disorders (e.g., Fanconi anemia).

Chromosome analyses can be done on *spontaneous abortion* specimens. Unless an inherited unbalanced translocation is demonstrated, however, the importance of a chromosome abnormality in an abortus is not known in terms of future pregnancies. In the face of three or more spontaneous abortions, it is best to study the parents' chromosomes. For a parent with mosaicism or a balanced translocation carrier state, the risks could be high, perhaps as much as 20% or even 100%. If delivery is premature or term and polysomy (e.g., trisomy 21) is found, there is a recurrence risk of about 1%. For the 45.X abnormalities (with or without mosaicism) and *de novo* translocations, additions, or deletions, the specific recurrence risks are unknown, but by analogy we would assume an increase. These facts, plus the knowledge that approximately 7% of stillbirths and neonatal deaths have an associated chromosome abnormality (25), make it worthwhile to karyotype all stillbirths and patients dying in the neonatal period. Among chromosomally abnormal patients who survive into infancy, childhood, or adulthood, there will be two modes of clinical presentation: 1. Mental retardation and/or associated minor or major congenital anomalies. 2. Disturbances of gonadal function or reproductive ability. For another small number of chromosome abnormalities, the clinical correlations are not clear (e.g., 47,XYY), or the abnormality is an incidental finding in a family evaluation or population survey.

The vast majority of the unbalanced autosomal abnormalities produce mental retardation and/or multisystem malformations; to a lesser degree this is true of X-chromosome polysomy. Some patients with the 45,X chromosome constitution will have obvious congenital anomalies but not mental retardation. Disturbances of gonadal function are usually manifested at about the time of expected puberty. This is the case with gynecomastia of

47,XXY and primary amenorrhea of 45,X or 45,X/46,XX. Balanced translocations are usually detected because of reproductive difficulties (infertility or miscarriages) or family investigations.

With the enhanced sensitivity of chromosome analysis we rely less on the ability to make a specific clinical diagnosis of a recognized syndrome. More and more chromosome analyses are being performed, with positive results, on the basis of general considerations. And it is in precisely such circumstances that new, often subtle, chromosome rearrangements are being demonstrated in the affected patient and a parent (Fig. 3-5). With a clinical diagnosis of Down syndrome, we expect to find an abnormality involving chromosome 21, while in nonspecific or new syndromes we are finding many new chromosome rearrangements that usually cannot be specified without the chromosome analysis. I believe it is these families that have gained the most assistance from cytogenetic advances. For without demonstration of the chromosome rearrangement, prenatal diagnosis might not be advised in subsequent pregnancies.

The environmental causes of acquired chromosome abnormalities and their consequences will be discussed in Chapter 7. But a few words about the relationship between acquired chromosome abnormalities and malignancies are in order here to emphasize that not all chromosome abnormalities are congenital in origin. Although there is probably some causal relation between chromosomal aberration and the development of malignancy, it is not clear which is the cause and which the effect. We have learned from the disorder xeroderma pigmentosum that changes in submicroscopic segments of the chromosomal material (mutations) can lead to cancer (26); so it seems likely that larger, microscopic changes could do the same. For this reason, several heritable disorders associated with an increased tendency toward both chromosomal rearrangement and the development of malignancies are being carefully investigated. These disorders include Fanconi anemia (27), Bloom syndrome (28), and ataxia telangiectasia (29). More common, however, are other neoplasias associated with more-or-less specific chromosome changes in the neoplastic cells, such as monosomy of chromosome 22 in meningiomas (30,31), loss of chromosome 7 or gain of chromosome 8 or 9 in myeloproliferative disorders (32), and development of the Philadelphia chromosome (22q−) in chronic myelogenous leukemia (33). Actually, the Philadelphia chromsome (Ph^1) results from a translocation of the distal long arm of a #22 to another chromosome, usually the terminal long arm of chromosome 9 (33). As the Ph^1 can be of diagnostic or even prognostic value (34), we anticipate that other chromosome aberrations which develop with neoplasia will have similar value, especially in determining the need for treatment or changes in treatment. Chromosome analysis of the malignant tissue, particularly bone marrow in leukemias, should become an increasingly important part of the management of such patients.

Clinical Disorders

A few of the more common chromosome disorders, including those listed in Table 3-6, will be discussed in this section, which is not intended to be anything more than a general conceptual guide.

The *Down syndrome* is relatively common, with a general incidence of approximately 1 in 1000. It can be associated with long life, though without mosaicism there are always significant stigmata, including moderately severe mental retardation (35,36, 37). The syndrome can result from more than one type of chromosome abnormality—trisomy and translocations—involving chromosome 21 (Figs. 3-4 and 3-6)(2). At least 1–2% of cases are literally inherited (translocations or parental mosaicism), and there is about a 1% empiric recurrence risk even when inheritance is not a factor. Moreover, there is an unequivocal increase in incidence with maternal age of 35 years or more at the time of delivery. The chronic nature of the disorder and its severity, commonness, recurrence likelihood, identifiable high-risk mothers, and possibility of prenatal diagnosis make Down syndrome uniquely important. Every health care worker should be able to recognize this disorder and understand that a chromosome diagnosis must always accompany the clinical diagnosis. As soon as the clinical diagnosis is entertained, a chromosome analysis must be done.

The clinical diagnosis is usually obvious at birth. The key features in a child include the distinctive facial appearance (Fig. 3-11) of upward slanting eyes, small, malformed ears, and large protruding tongue; other superficial features include poor muscle tone, short stubby hands with incurved fifth fingers (clinodactyly), and abnormal dermatoglyphics including simian creases and arch tibial patterns on the balls of the feet. Major visceral anomalies involving the heart, gastrointestinal tract, and urinary

TABLE 3-6
The Most Common Specific Congenital Chromosomal Disorders

Down syndrome	Turner syndrome
Edwards syndrome	Klinefelter syndrome
Patau syndrome	Triplo-X constitution
Warkany syndrome	47,XYY constitution

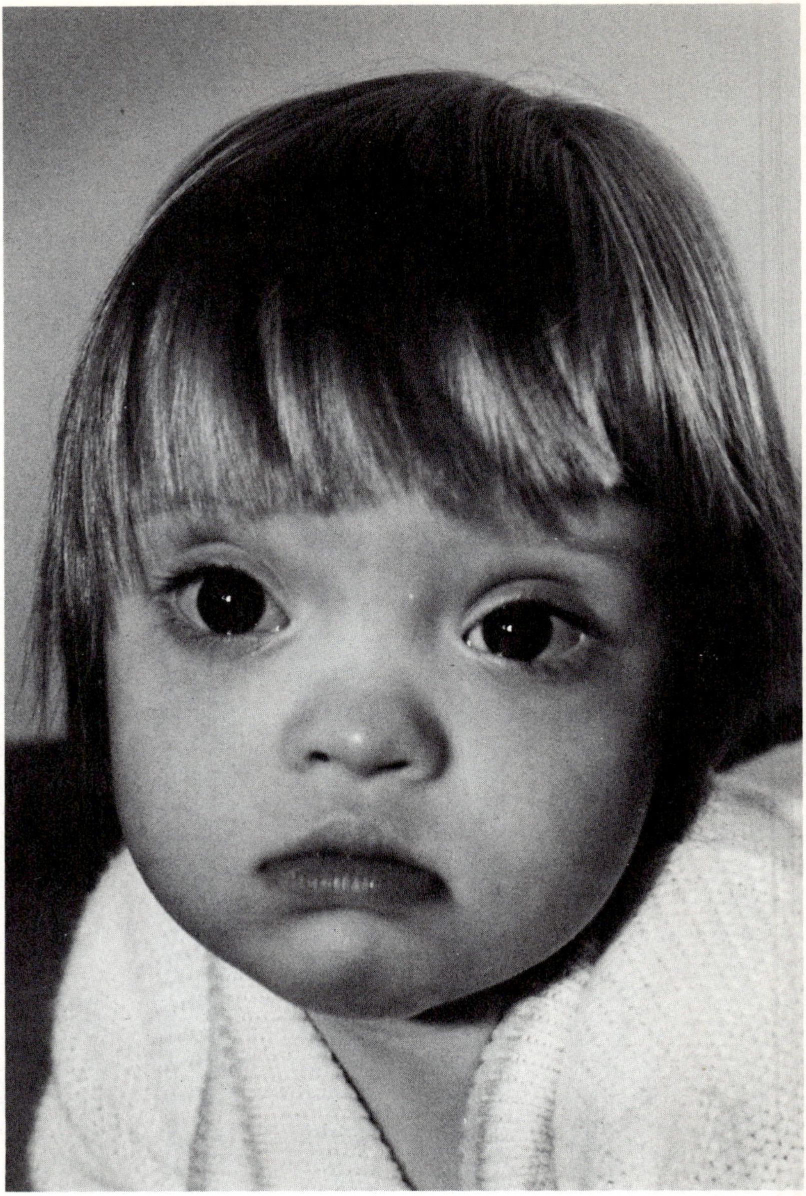

Figure 3–11 Facies of a young child with the Down syndrome. Note the round face, wide base of the nose (telecanthus), and down-curved mouth.

system are not uncommon; and usually the patient's longevity is decreased. Thyroid dysfunction (increased or decreased) is common, as are psoriasis, gout, and presenile dementia. Procreation is unknown for affected men without mosaicism, but is reported for women, with a 50% risk of having a similarly affected child.

This disorder is one of, if not *the* most common mental retardation syndrome having a specifiable cause among patients in state and municipal institutions for the retarded. As such, it serves as its own argument for state and municipal support of programs that would minimize the frequency of this disorder (see Chapter 15). Very few states, however, have supported a preventive effort.

The *Patau syndrome* is a more devastating disorder (38). In the absence of mosaicism, death usually occurs in early infancy. The associated anomalies include microcephaly, microphthalmos, cleft lip and palate, congenital heart disease, polycystic kidneys, and polydactyly. The grotesque appearance usually makes an indelible impression on the parents. This syndrome has an incidence of about 1 in 6000, and, like the Down syndrome, can have more than one chromosomal basis: trisomy 13 and Robertsonian translocation, the latter sometimes being heritable. In addition, there are numerous instances of the partial syndrome resulting from the extra presence of only parts of chromosome 13. In some cases there will be only 46 chromosomes with one bearing material reciprocally translocated from the long arm of a #13. In other cases there are 47 chromosomes, the extra one being a #13 with a deleted long arm (47,+13q−). In still other cases there is mosaicism (46/47,+13). Also, the features of the full clinical syndrome overlap considerably with the autosomal recessive disorder known as the Meckel syndrome, which would have a 25% recurrence risk (see Chapter 4). The point is that with both a variable clinical picture and multiple chromosomal bases, it is imperative to obtain a chromosome analysis—including postmortem—if the clinical diagnosis of Patau syndrome is entertained even lightly. Parental chromosome analyses would also be necessary if the affected child's karyotype showed a Robertsonian translocation, reciprocal translocation, or an extra deleted chromosome. Such analyses would be necessary even if the parents have no plans for more children, since previous siblings may be balanced carriers. Trisomy 13, like trisomy 21, is more likely to occur with maternal age of 35 years or more.

The *Edwards syndrome* is as severe as the Patau syndrome and clinically is sometimes confused with it (38). It is usually associated with trisomy 18, and has an incidence of about 1 in 8000 and a positive correlation with increased maternal age (35 years and older). The main features are microcephaly, prominent occiput, small eyes, misshapen ears, prominent tip of the nose, small mouth, micrognathia, short sternum, congenital heart disease, hypoplastic labia in females, overriding clenched fingers and rocker-bottom feet. Death usually occurs in early infancy. Although the vast majority of cases with this syndrome have full trisomy 18, there are known instances of unbalanced reciprocal translocations, mosaicism, and even normal karyotypes. Moreover, if this clinical diagnosis is made and the chromosome constitution is normal, then the patient may have the "pseudotrisomy 18 syndrome," an autosomal recessive trait with a 25% recurrence risk. Again, it is critical to analyze the patient's chromosomes, postmortem if necessary. As with Down and Patau syndromes, any of the chromosomally abnormal states warrant amniocentesis for prenatal diagnosis in subsequent pregnancies.

The *Warkany syndrome* was first recognized in 1962, but since then increasing numbers of patients have been diagnosed. This syndrome exemplifies the usefulness of eponyms for designating chromosomal disorders. Ordinarily, the Warkany syndrome is associated with trisomy-8 mosaicism (46/47,+8) (Fig. 3-12), but it

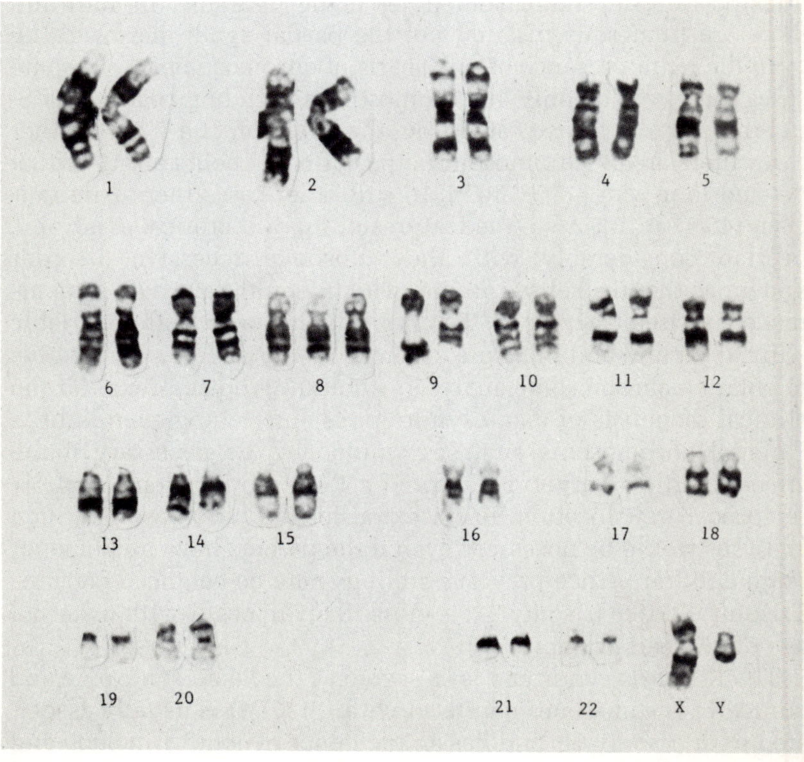

Figure 3–12 Trisomy 8 karyotype from a patient with the Warkany syndrome. There are three #8 chromosomes. The long secondary constriction in one #9 (9qh +) is a normal variant. (From Atkins, Holmes, Riccardi, *J. Pediatr* 84:302, 1974. With permission.)

is also known to occur with reciprocal translocations (39). Moreover, some patients with trisomy 8 mosaicism show no evidence of the syndrome. And finally, bone marrow trisomy 8 mosaicism, presumably acquired, has been found in association with a number of myeloproliferative disorders. Its incidence is not known, but it is probably more common than is usually thought. This is explained at least in part by the mosaicism and the tendency for the trisomic cell line to be lost with increasing age of the patient. The features of the Warkany syndrome include mental retardation, typical facial appearance (Figs. 3-13, 3-14), ocular defects, deep plantar and palmar skin furrows, congenital contractures about the major joints, camptodactyly, abnormal vertebrae, absent patellae, and congenital hydronephrosis. Because of possible mosaicism, the diagnosis cannot be ruled out unless chromosome analysis has been performed on cells cultured from a skin biopsy. At least three patients are known to have had mosaic trisomy 8 in lymphocytes or skin without the congenital syndromes. One was a sibling to Warkany's patient (19); one was a normal woman except for recurrent miscarriage (40); and one presented with acute leukemia (41). Thus, to call this the "trisomy 8 syndrome," with or without the qualifier of mosaicism (42,43,44), would overlook the fact that it may result from a translocation (without mosaicism) and that mosaic trisomy 8 is not always associated with the syndrome. This syndrome also highlights the role of skin biopsy in clinical cytogenetic diagnosis and the need to study parents and siblings of affected persons in at least selected instances.

Except for *Trisomy 22* (45), almost all other clinically distinctive autosomal abnormalities are due to partial trisomies or deletions (46), often associated with a balanced reciprocal translocation in one of the parents. Among the deletion disorders, those best known are the cri-du-chat syndrome (5p−), the Wolfe syndrome (4p−), and the 18q− syndrome. Another one will be described here because it exemplifies the group and illustrates other points as well. Deletions of 11p have been characterized by mental retardation, aniridia, urinary tract anomalies, and pseudohermaphroditism (Fig. 3-15). Three patients with this disorder are known to the author (5), and they are instructive in at least three ways: 1. The two boys have severe mental retardation, aniridia, severe hypospadias and hydronephrosis secondary to ureteral anomalies. Early chromosome analyses without banding techniques yielded "normal" karyotypes. With G-banding, both showed the deletion. 2. The girl, with severe mental retardation, aniridia, ambiguous genitalia, and a Wilms tumor, had an XY sex chromosome constitution. Her chromosome 11 short-arm deletion demonstrates that autosomal aberrations have to be considered in evaluating ambiguous genitalia and pseudohermaphroditism. 3.

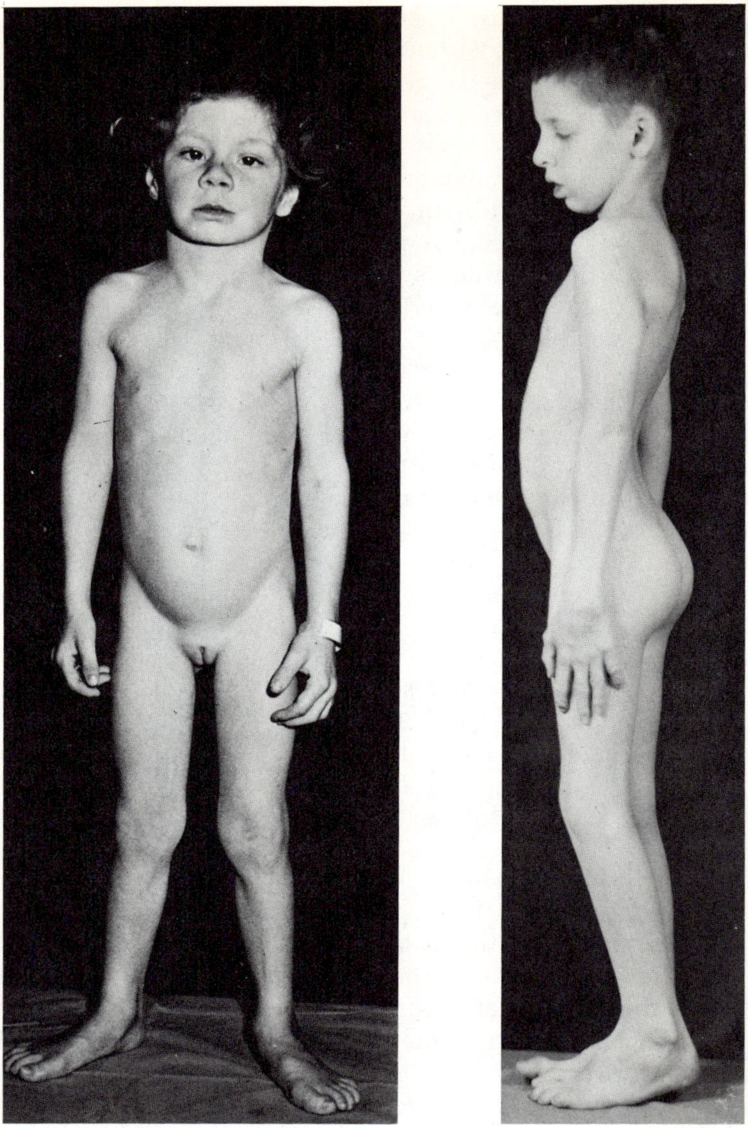

(a) (b)

Figure 3–13 The appearances of two young patients who have the Warkany syndrome. (a) This 9-year-old girl shows the facies typical of the syndrome; widespread nipples, slight pectus excavatum, excessively flexed fingers, genu valgum, pes planus, hammer-toe deformities, and lateral deviation of the great toes. (b) This 10-year-old boy's stance reflects the joint contractures in this disorder: There is fixed flexion of the left hip and knee, moderate plantar flexion and internal rotation at the ankles, and exagerated lumbar lordosis. (From Riccardi, Atkins, Holmes, *J Pediatr* 77:664, 1970. With permission.)

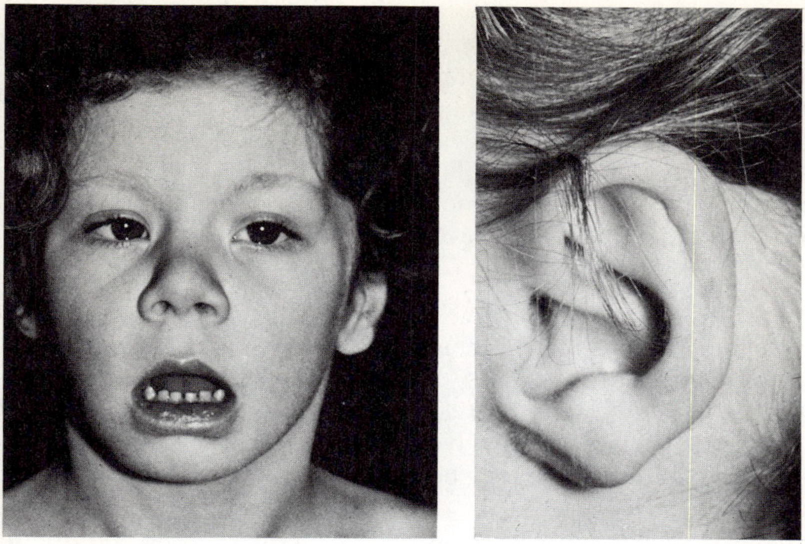

(a) (b)

Figure 3–14 Facies and ear morphology of a patient with the Warkany syndrome [Case (a), Fig. 3–13]. (a) The nasal root is broad, giving the eyes a wide-set appearance (telecanthus), but there is no hypertelorism. The right eye shows a convergent strabismus; the nasal tip is upturned and prominent; the mouth is held open; the lower jaw projects forward; and the teeth are widely spaced. (b) Both pinnae show an unusual configuration: superiorly the helix is thick; in the midportion the helix abruptly flattens, with loss of the scaphoid fossa; inferiorly, the incisura is broad and the lobule is prominent with an anterior groove. (From Riccardi, Atkins, Holmes, *J Pediatr* 77:664, 1970. With permission.)

Aniridia, combined with other anomalies, especially pseudohermaphroditism, has been strongly associated with the development of a Wilms tumor. Thus, it becomes important to obtain chromosome analyses on patients with aniridia that is not part of the autosomal dominant disorder, as well as patients presenting with Wilms tumors, to further characterize the positive association of this chromosome aberration (11p−) with the development of the tumor.

A more well known though rare disorder will serve to make one final point about evaluations of families that have children with structural chromosome abnormalities, including deletions. A young boy had a chromosome analysis because of developmental delay, short stature, facial assymetry, auricle abnormalities, and other anomalies. All cells analyzed showed a terminal deletion of a chromosome 18 long arm. He was the first-born of young parents. Their chromosomes were analyzed, revealing a balanced reciprocal translocation involving the long arms of chromosomes 16 and 18 in the father. The couple felt that this knowledge, because of the availability of prenatal diagnosis, was of immeasurable value for their future family planning. Similar case reports for many other structural chromosome abnormalities, including the 4p−, 5p−, and 47,+9q− syndromes, have been published (46).

The need to make a specific cytogenetic diagnosis of *sex chromosome disorders* is based on the principles that apply to autosomal abnormalities and on several additional factors. These added considerations bear directly on long-term patient management.

1. Mosaicism is common, especially with monosomy-X.
2. Two or more chromosome abnormalities may account for a given clinical disorder [i.e., 46,XXiq (long-arm isochromosome) versus 46,XXp − (short-arm deletion) versus 45,X].
3. Polysomy may involve both sex chromosomes (e.g., 48,XXYY) and/or there may be more than one extra chromosome (e.g., 49,XXXXY).
4. If a 45,X cell line is mosaic with a 46,XY cell line, the risk of gonadal malignancy is increased, and the Y chromosome's pres-

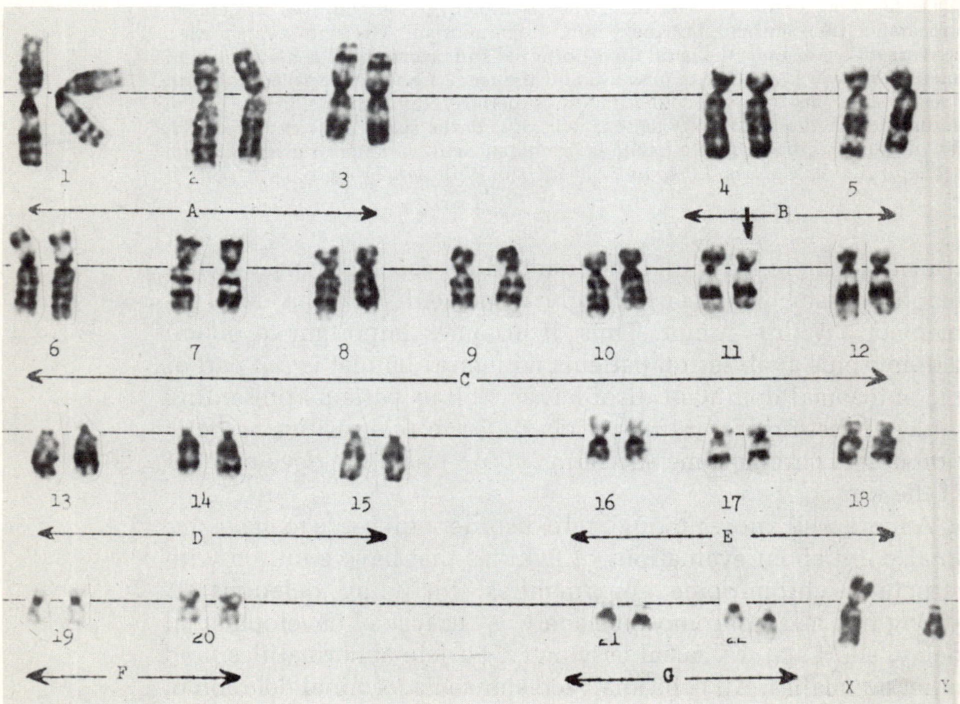

Figure 3–15 InterstitialTerminal deletion of the short arm of a chromosome #11 46,XY, Del(11)(p12;p15). The deletion is subtle, and had been missed in a nonbanded karyotype.

ence cannot be ruled out by buccal smears with fluorescent stains (47).

5. Nonchromosomal disorders may clinically overlap with sex chromosome disorders (i.e., familial gonadal dysgenesis versus mixed gonadal dysgenesis due to a 45,X/46,XY chromosome constitution).

6. The clinical stigmata may be few and subtle, or incompletely defined (e.g., 47,XXX and 47,XYY).

7. The characteristic clinical features will vary according to the patient's age (see Table 3-7). The importance of this has been minimized, if not totally obfuscated, by the widespread misapplication of the eponyms "Turner syndrome" and "Klinefelter syndrome." These terms actually designate postpubertal clinical abnormalities, including hypogonadism and infertility. Nonetheless, they are usually applied when certain chromosome abnormalities are shown (45,X and 47,XXY), regardless of the age and physical features of the patients, and this is inappropriate (48). Each type of diagnosis (clinical and chromosomal) must be established separately, by distinct criteria.

If it is appreciated that chromosome disorders in general may present at different ages and with varying features, more accurate clinical and chromosomal diagnoses can be made. And for the sex chromosome abnormalities, where a clinical diagnosis often is not apparent at birth or in early childhood, this consideration is most important. Because such abnormalities in many infants and young children escape clinical detection, newborn cytologic screening programs are the most effective way to make an early diagnosis (49,50). However, controversy over the medical, social, legal, and ethical impact of premorbid diagnosis of sex chromo-

TABLE 3-7
Indications for Buccal Smear as a Function of Age

Newborn, infancy	**Adolescence, puberty**
Pedal edema, webbed neck, with or without aortic coarctation, or congenital heart disease of other types.	Delayed or incomplete development of secondary sex characteristics
Ambiguous genitalia	Small genitals
Female with "inguinal hernias"	Primary amenorrhea
	Gynecomastia
Childhood	**Adulthood**
Persistent cryptorchidism	Infertility
Small testes	Incomplete sexual development or performance
Short stature	Persistent acne, combined with tall stature and agressive personality
Developmental delay, mental retardation	Suspicion of a specific syndrome diagnosis (e.g., Klinefelter or Turner syndromes).
Radioulnar synostosis	

some aberrations has essentially terminated such efforts (51). This will retard our understanding of the natural history of these disorders, and it means that the burden of early diagnosis, for application of secondary and tertiary preventive efforts, falls to the clinician.

Sex chromosome abnormalities are relatively common (52). They account for at least 25% of spontaneous abortions (45,X) and have a combined frequency in newborns of about 1 in 500. They are manifested sooner or later by any one or combination of the following features: ambiguous or otherwise abnormal genitalia, hypogonadism, amenorrhea, infertility, congenital heart disease, urinary tract anomalies, very tall or very short stature, other skeletal abnormalities, abnormal dermatoglyphic patterns, mental retardation, and perhaps pathologically aggressive personalities.

The *45,X chromosome constitution* is present in approximately 2% of all conceptions, 25% of spontaneous abortions, and 1 in 2500 liveborn females. Physical stigmata among the liveborns are varied, ranging from essentially none to include abnormal dermatoglyphics, redundant cervical skin folds (webbing), nonpitting edema of the dorsum of the hands and feet, urinary tract anomalies, congenital heart disease (especially coarctation of the aorta), shortened fourth and fifth metacarpals, and occasionally abnormal genitalia (53,54). Streak ovaries (gonadal dysgenesis) are virtually always present, eventually accounting for hypogonadism and infertility (Turner syndrome). When there is mosaicism with a normal female cell line (45,X/46,XX) many or most of these features may be absent, the person being essentially normal except perhaps for menstrual irregularities and ovarian abnormalities. When there is mosaicism with a 46,XY cell line (45,X/46,XY), mixed gonadal dysgenesis often results; ambiguous genitalia, usually appearing to represent an incompletely masculinized male, may be present. Since neither gonad is expected to function as a testis, Y-bearing nonscrotal gonads should be removed before development of neoplasia, usually a gonadoblastoma or dysgerminoma. Thus, one should consider 45,X/46,XY mosaicism whenever the buccal smear is negative for both X-chromatin and Y-chromatin and the clinical evaluation reveals ambiguous genitalia. The long-term importance of recognizing the various possible types of 45,X mosaicism cannot be overemphasized. The same types and combinations of clinical problems are also caused by structural X-chromosome aberrations: the net effect of these aberrations is to render the person monosomic for various portions of the X-chromosome. They include ring X-chromosomes, isochromosomes, and deleted X-chromosomes. Moreover, since one of the two X-chromosomes is heterochromatic in interphase (Barr body) and since an abnormal X-chromosome is the more usual one to become heterochromatic, an unusually small or large Barr body may indicate such a structural aberration, i.e., smaller with a ring X, short-arm isochromsome or deletion, and larger with a long-arm isochromosome. Thus, the presence of Barr bodies in a buccal smear does not rule out monosomy for at least part of the X-chromosome.

The point here is that assistance from a clinical geneticist is appropriate in all suspected and tentative cases of complete or partial monosomy for the X-chromosome, with or without mosaicism.

The term "monosomy-X," with or without the additional qualifications "partial" or "mosaic" is the preferred way to refer to the chromosome abnormalities described in the preceding paragraph and to the clinical disorders associated with them. The term "Turner syndrome" should be reserved for those cases where there is sexual infantilism (hypogonadism), amenorrhea, short stature, webbed neck, increased carrying angles, and shortened fourth and fifth metacarpals. Since some of these features cannot be demonstrated before puberty, the eponym, strictly speaking, should not be used until expected pubarche. Mental retardation is not a feature of this disorder. When present, it may reflect an independent etiology (i.e., hemizygosity for an X-linked mental retardation gene) or a secondary handicap (i.e., hearing deficiency secondary to the recurrent otitis media that is common with monosomy-X). The phrase "male Turner syndrome," though still used occasionally, is a misnomer. The nosologic entity of Noonan syndrome accounts for most male patients whose features overlap with true Turner syndrome (55,56); for the remainder other specific or descriptive diagnoses would apply.

The chromosome constitution *47,XXX* was originally thought to have little if any clinical consequence. However, it has become increasingly apparent that mental retardation and a variety of minor and major congenital malformations can result from this aberration (57). Fertility is not necessarily impaired and offspring are at risk for X-chromosome polysomy. The tetra-X (48,XXXX) and penta-X (49,XXXXX) abnormalities are associated with more obvious and severe clinical problems.

The *47,XXY* chromosome constitution is associated with different key features at different ages, and the same features may also result from other X-chromosome abnormalities or may have an entirely different etiology. At birth there may only be abnormal dermatoglyphics and/or cryptorchidism. In childhood, additional features may include mental retardation or learning disabilities, radioulnar synostosis, tall stature, and small testes. At or beyond puberty, which may be delayed, these features can be anticipated: gynecomastia, small penis, feminine pattern of obesity, and general signs of hypogonadism. Without mosaicism, infertility is uniformly present. Mental retardation is very variable in severity and may be mild or absent. Personality disorders, including excessive aggressiveness, may also be present. Early clarification of the correct cytogenetic diagnosis is important from at least two viewpoints: 1. Many if not all of the associated features, whether alone or in combination, may have other causes, including other chromosome aberrations (e.g., 48,XXYY or 48,XXXY). 2. Realistic expectations can be established and proper assistance provided at optimal times. Frequently, an adolescent boy shows what may be the secondary consequences of frustration from unappreciated learning problems and chastisement for his delayed and/or incomplete puberty. If educational assistance and hormonal therapy (and even mastectomy for gynecomastia) were instituted early, the secondary consequences might be alleviated. Cytogenetic diagnosis should include both a buccal smear and karyotyping to consider the presence and extent of mosaicism with a 46,XY cell line. Such mosaicism could markedly temper the severity of the syndrome and even allow for fertility. The 47,XXY chromosome constitution has an incidence between 0.5 and 2 per 1000 newborn males (49,58). 3. Other sex chromosome aneuploidy or autosomal aneuploidy may occur in first- and second-degree relatives (59). Thus, parents of a young child with a 47,XXY karyotype should be afforded amniocentesis in subsequent pregnancies.

The term "Klinefelter" syndrome is often applied whenever there are extra X-chromosomes in cells containing at least one Y-chromosome (53). These karyotype variants include 48,XXYY, 48,XXXY, 49,XXXYY, and 49,XXXXY. Generally, the larger the number of extra X-chromosomes, the more severe the disorder. The 49,XXXXY abnormality, for example, is virtually always associated with severe mental retardation (60).

A *47,XYY* chromosome constitution also has an incidence of about 1 in 900 male newborns; although the consequences of this abnormality are not entirely clear, they have been the subject of intense debate (61,62). It would seem that certain types of aggressive, antisocial behavior may be associated with this chromosome aberration (61). In some instances frank mental retardation may occur. Tall stature and moderate or severe facial acne are less controversial clinical features.

In closing this chapter, it is fitting to reemphasize one of the most important new aspects of clinical cytogenetics. *Acquired chromosome abnormalities* include:

1. Those that are not apparent at birth and are detected in association with another genetic disorder (e.g., Fanconi anemia).
2. Those due to exposure to environmental agents (e.g., irradiation).
3. Those associated with the development of a malignancy.

The third group deserves particular emphasis here since this phenomenon is being observed increasingly often and probably will soon play a role in the routine diagnosis and management of many cancer patients. The Philadelphia chromosome (Ph^1) has been recognized in conjunction with chronic myelogenous leukemia (CML) for years. Although its precise relationship to the malignant process is not yet clear, its presence in the malignant cells themselves suggests that it is more than coincidence. It occurs in about 85% of CML patients, and its absence or multiple presence is associated with a more grave prognosis (34). The Ph^1 chromosome actually is a chromosome 22 whose distal long arm has been translocated to another chromosome, most often the distal long arm of a #9 (Fig. 3-16). Other myeloproliferative disorders have been associated with aneuploidy for C-group (numbers 6–12) chromosomes, including monosomy for #7. Chromsome-8 trisomy has been demonstrated in bone marrow cells in a variety of such disorders, including polycythemia vera rubra, myelofibrosis, CML, and especially acute myelogenous leukemia (7,32). Although we cannot yet draw definite conclusions regarding prognosis and course of the disease from the

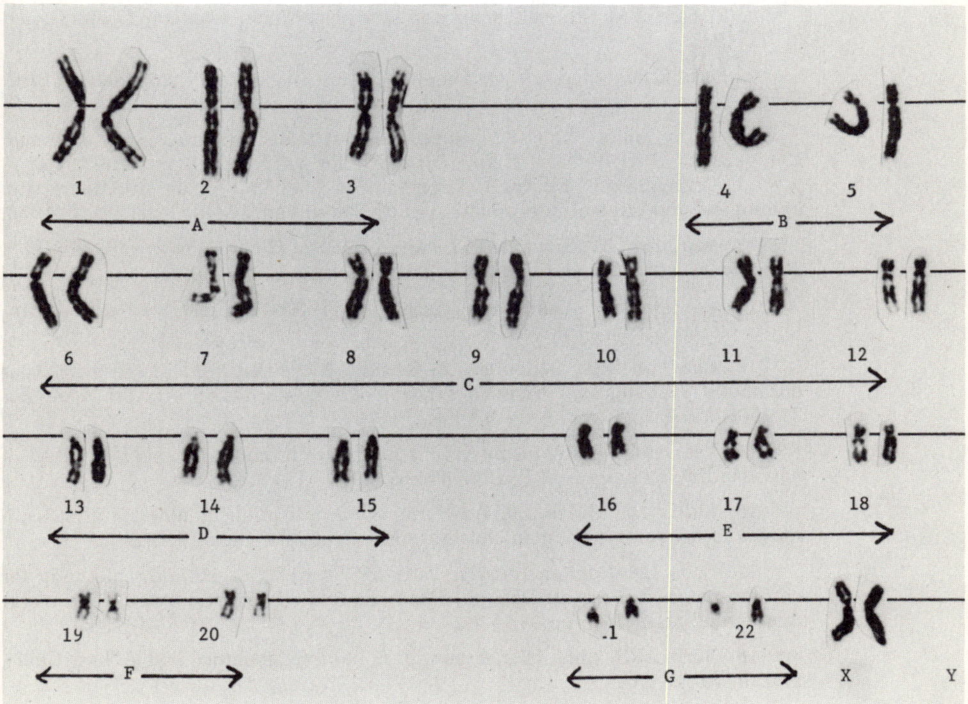

Figure 3–16 Karyotype of a peripheral blood cell from a patient with chronic myelogenous leukemia. A Philadelphia chromosome (deleted #22) is seen, with the deleted portion translocated to the end of a 9q.

presence of an extra chromosome 8, in the long run this should be possible.

Cited References

1. Lewandowski RC, Yunis JJ: New chromosomal syndromes. *Am J Dis Child* 129:515, 1975.

2. Williams JD, Summit RL, Martens PR, et al: Familial Down syndrome due to t(10;21) translocation: Evidence that the Down phenotype is related to trisomy of a specific segment of chromosome 21. *Am J Hum Genet* 27:478, 1975.

3. Smith DW: *Recognizable Patterns of Human Malformations* (2nd ed). Philadelphia, WB Saunders, 1976.

4. Holmes LB, Moser HW, Halldorsson S, et al: *Mental Retardation—An Atlas of Diseases with Associated Physical Abnormalities.* New York, Macmillan, 1972.

5. Riccardi VM, Sujansky E, Smith ACM: Aniridia, mental retardation, genital anomalies, and Wilms' tumor in patients with 46,XY,11p−. (unpublished observations).

6. Riccardi VM, Crawford JD, Cohen J, et al: Cytogenetic implications of low X-chromatin frequency in congenital adrenal hyperplasia. *Lancet* 1:497, 1972.

7. Hsu LYF, Alter AV, Hirschhorn K: Trisomy 8 in bone marrow cells of patients with polycythemia vera and myelogenous leukemia. *Clin Genet* 6:258, 1974.

8. Riccardi VM: Clinical genetics and the obstetrician. *Contemp Obstet Gynec* 2:105, 1973.

9. Paris Conference (1971): Standardization in Human Cytogenetics. *Birth Defects Original Article Series,* vol VIII, No 7, pp 1–46, 1972.

10. Paris Conference (1971), Supplement (1975): Standardization in Human Cytogenetics. *Birth Defects Original Article Series,* vol XI, No 9, pp 1–36, 1975.

11. Gripenberg U, Elfving J, Gripenberg L: A 45,XX,−21 child: Attempt at a cytological and clinical interpretation of the karyotype. *J Med Genet* 9:110, 1972.

12. Yeatman GW, Riccardi VM: Partial trisomy of chromosome 14; (+14q−). *Birth Defects Original Article Series,* vol XII, No 5, p. 119, 1976.

13. Wertelecki W, Graham JM, Sergovich FR: The clinical syndrome of triploidy. *Obstet Gynec* 47:69, 1976.

14. Chaganti RSK, Schonberg S, German J: A many fold increase in sister chromatid exchanges in Bloom's syndrome lymphocytes. *Proc Natl Acad Sci* 71:4508, 1974.

15. Peakman DC, Robinson A: Indicators for chromosome analysis: Effect of new banding techniques. *J Pediatr* 88:166, 1976.

16. Miller OJ, Miller DA, Warburton D: Application of new staining techniques to the study of human chromosomes. *Prog Med Genet* 9:1, 1973.

17. Lubs HA, McKenzie WH, Patil SR, et al: New staining methods for chromosomes. in Prescott DM (ed): *Methods in Cell Biology,* vol 6, New York, Academic Press, 1973, pp 345–380.

18. Verma RS, Lubs HA: A simple R banding technic. *Am J Hum Genet* 27:110, 1975.

19. Warkany J, Rubinstein JH, Soukup SW, et al: Mental retardation, absence of patellae, other malformations with chromosomal mosaicism. *J Pediar* 61:803, 1962.

20. Riccardi VM: Trisomy 8: An international study of 70 patients. *Birth Defects Original Article Series* (in press).

21. Borgaonkar DS, Bolling DR: International registry of abnormal karyotypes. *Humangenetik* 28:317, 1975.

22. Boue A, Boue J: Chromsome abnormalities and abortion. in Continho EM, Fuchs F, (eds): *Physiology and Genetics of Reproduction,* Part B, New York, Plenum Press, 1974, pp. 323–371.

23. McConnel HD, Carr DH: Recent advances in the cytogenetic study of human spontaneous abortion. *Obstet Gynec* 45:547, 1975.

24. Miller OJ, Breg WR: Current concepts in genetics: Autosomal chromosome disorders and variations. *New Engl J Med* 294:596, 1976.

25. Kuleshov NP: Chromosome abnormalities of infants dying during the perinatal period and premature newborn. *Hum Genet* 31:151, 1976.

26. Cleaver JE, Bootsma D: Xeroderma pigmentosa: Biochemical and genetic characteristics. *Ann Rev Genet* 9:19, 1975.

27. Swift MR, Hirschhorn K: Fanconi's anemia: Inherited susceptibility to chromosome breakage in various tissues. *Ann Intern Med* 65:496, 1966.

28. German J: Bloom's syndrome: I. Genetical and clinical observations in the first twenty seven patients. *Am J Hum Genet* 21:196, 1969.

29. Hecht F, McCaw BK, Koler RD: Ataxia telangiectasia-clonal growth of translocation lymphocytes. *New Engl J Med* 289:286, 1973.

30. Porter IH, Paul B: Chromosome changes and the supression of malignancy in a human meningioma. *Am J Hum Genet* 27:72A, 1975.

31. Zankl H, Seidel H, Zang KD: Sex chromosome loss in human tumors. *Lancet* 1:221, 1975.

32. Rowley JD: Nonrandom chromosomal abnormalities in hematologic disorders of man. *Proc Nat Acad Sci* 72:152, 1975.

33. Rowley JD: A new consistent chromosomal abnormality in chronic myelogenous leukemia identified by quinacrine fluorescence and Giemsa staining. *Nature* 243:290, 1973.

34. Sakurai M, Hayata I, Sandberg AA: Prognostic value of chromosomal findings in Ph-positive chronic myelocytic leukemia. *Cancer Res* 36:313, 1976.

35. Penrose LS, Smith GF: *Down's Anomaly.* Boston, Little, Brown, 1966.

36. Gayton WF, Walker L: Down syndrome: Informing the parents. *Am J Dis Child* 127:510, 1974.

37. Center for Disease Control: Incidence of Down syndrome in the United States. Congenital Malformation Surveillance, July 1974–June 1975, issued December 1975, Atlanta, pp 7–9.

38. Taylor AI: Autosomal trisomy syndromes: A detailed study of 27 cases of Edwards' syndrome and 27 cases of Patau's syndrome. *J Med Genet* 5:227, 1968.

39. Sanchez O, Yunis JJ: Partial trisomy 8 (8q24) and the trisomy 8 syndrome. *Humangenetik* 23:297, 1974.

40. Caspersson T, Lindsten J, Zech L, et al: Four patients with trisomy 8 identified by fluorescence and giemsa banding techniques. *J Med Genet* 9:1, 1972.

41. Riccardi VM: T8 Mosaicism in the skin of a patient with leukemia. *Birth Defects Original Article Series,* vol XII, No 1, p 187, 1976.

42. Riccardi VM, Atkins L, Holmes LB: Absent patellae, mild mental retardation, skeletal and genitourinary anomalies, and C group autosomal mosaicism. *J Pediatr* 77:664, 1970.

43. Atkins L, Holmes LB, Riccardi VM: Trisomy 8. *J Pediatr* 84:302, 1974.

44. Cassidy SB, McGee BJ, Van Eyp J, et al: Trisomy 8 syndrome. *Pediatrics* 56:826, 1975.

45. Uchida IA, Byrnes EM: Confirmation of Trisomy 22 with fluorescence banding. *Am J Hum Genet* 28:189, 1976.

46. Centerwall WR, Beatty-DeSana JW: The trisomy 9p syndrome. *Pediatrics* 56:748, 1975.

47. Winters AJ, Benirschke K, Whalley P, et al: Mosaicism and lack of fluorescence of Y chromsome. *Obstet Gynec* 46:367, 1975.

48. Robinson A, Puck T, Tennes K: The 47,XXX karotype. *Lancet* 1:1343, 1974.

49. Goad WB, Robinson A, Puck TT: Incidence of aneuploidy in a human population. *Am J Hum Genet* 28:62, 1976.

50. Robinson A: Neonatal deaths and sex-chromosome anomalies. *Lancet* 1:1223, 1974.

51. Culliton BJ: Patients' rights: Harvard is site of battle over X and Y chromosomes. *Science* 186:715, 1974.

52. Gerald PS: Current Concepts in Genetics: Sex chromosome disorders. *New Engl J Med* 294:706, 1976.

53. Federmann DD: *Abnormal Sexual Development: A Genetic and Endocrine Approach to Differential Diagnosis.* Philadelphia, WB Saunders, 1967.

54. Engel E, Forbes AP: Cytogenetic & clinical findings in 88 patients with congenitally defective or absent ovaries. *Medicine* 44:135, 1965.

55. Bolton MR, Pugh DM, Mattioli LF, et al: The Noonan syndrome: A family study. *Ann Intern Med* 80:626, 1974.

56. Turner G, Collins E: The Noonan syndrome—A review of the clinical and genetic features of 27 cases. *J Pediatr* 83:941, 1973.

57. Tennes K, Puck M, Bryant K, et al: A developmental study of girls with trisomy X. *Am J Hum Genet* 27:71, 1975.

58. Witkin A, Mednick SA, Schulsinger F, et al: Criminality in XYY and XXY men. *Science* 193:547, 1976.

59. Dumars KW, Fialko G, Willis J: Prenatal diagnosis of XYY karyotype in a family previously delivered of a child with D trisomy. *Clin Res* 24:170A, 1975.

60. Hayek A, Riccardi VM, Atkins L, et al: 49,XXXXY chromosomal anomaly in a newborn. *J Med Genet* 8:220, 1971.

61. Hamerton JL: Human population cytogenetics: Dilemmas and problems. *Am J Hum Genet* 28:107, 1976.

62. Borgaonkar DS, Shah SA: The XYY chromsomal male—or syndrome. *Progr Med Genet* 10:135, 1974.

General References

A. Hamerton J: *Human Cytogenetics* vol I; *General Cytogenetics* vol II; *Clinical Cytogenetics.* New York, Academic Press, 1971.

B. Kim HJ, Hsu LYF, Paciuc S, et al: Cytogenetics of fetal wastage. *New Engl J Med* 293:844, 1975.

C. Ohno S: *Sex Chromosomes and Sex-linked Genes.* Berlin, Springer-Verlag, 1967.

D. Smith DW, Wilson AC: *The Child with Down's Syndrome (mongolism).* Philadelphia, WB Saunders, 1973.

CHAPTER FOUR

GENETIC MECHANISMS OF DISEASE – MENDELIAN DISORDERS

Introduction

The predictability of genetic disorders is one of their key features. Understanding the risks for their occurrence and recurrence involves five sets of considerations:

1. The *interplay of environmental factors* (including social ones) *with genetic factors* determines a disorder's onset, its clinical course, and ultimate impact. An individual's genotype is not the sole determinant in its final expression. This is intuitively obvious for some things (i.e., death from an accident before Huntington chorea is manifest) and will be dealt with extensively in the discussion of pharmacogenetics (Chapter 14).

2. It is also recognized that *certain genes will modify or limit the expression of other genes*. Thus, for example, in a genotypic (karyotypic) male the virilizing effect of the adrenogenital syndrome II (21-betahydroxylase deficiency) will be less important than in a female. Likewise, the sickling phenomenon of sickle-cell anemia would be sharply diminished if the person were also hemizygous or homozygous for glucose-6-phosphate dehydrogenase deficiency.

3. *Empiric data* may be the key to ascribing risk figures. This point has already been brought up several times and will be amplified below.

4. The principles of *gene segregation* and *reassortment* as delineated by Mendel allow for the accurate forecast of recurrence risks among various family members once the details of each case have been defined.

5. Defining those details, a task of the clinical geneticist, involves another set of principles relating to *genotype* and *phenotype*. These principles account for occasional exceptional features of disorders and encompass such phenomena as spontaneous mutations, genetic heterogeneity, and phenocopies.

For many clinicians, genetics is an esoteric discipline, steeped in its own jargon and rituals and not always in touch with the real concerns of the affected patient or family. A consideration of the above concepts in clinical terms should help undercut this view. The purpose of this chapter is to show how Mendelian principles can be applied in clinical practice when the situation is straightforward, how to recognize when it is not straightforward, and what else must then be considered to clarify the situation.

Principles, Terminology, and Techniques

As we have seen, the prevention of genetic disorders involves the use of both empiric data and genetic principles. In Chapter 3 it was shown that *empiric data* define at-risk groups (e.g., those most likely to have a chromosome abnormality) or recurrence risks (e.g., the 1% risk of another offspring having aneuploidy following the birth of a child with trisomy 21) and that *genetic principles* dictate both a chromosomal evaluation of parents of a child with a chromosome rearrangement and a relatively high recurrence risk when a carrier parent had been identified. The combined use of empiric and logical (Mendelian genetic) data is the foundation of efforts to assist families at risk for the occurrence or recurrence of Mendelian disorders (1). Only after the correct, most specific diagnosis has been made, can it be decided whether to emphasize empiric or genetic considerations. Figures 4-1, 4-2, and 4-3 depict the relationship of recurrence risks to various types of empirically derived data. When either a unilateral or bilateral retinoblastoma is known to be inherited, mainly Mendelian principles apply (i.e., 50% transmission of the mutant gene); while with isolated cases one relies on empiric data.

This need to coordinate genetic and empiric factors is further exemplified by phenylketonuria (PKU), a disorder known to almost everyone. If the eighth child in a family is found to have PKU, the genetic conclusions include a 1 in 4 (25%) recurrence risk for subsequent siblings since PKU is an autosomal recessive disorder. In addition, we know from empiric data, gathered from previously affected families, that each previous sibling may be affected subclinically. Thus, each one must be checked for chemical evidence of this disorder, especially females since they would have a substantial risk of bearing a child with microcephaly and mental retardation.

Some human features or traits have polygenic determinants, a topic to be covered in the subsequent chapter. With many other traits, especially pathologic ones, only a pair of genes for each characteristic is responsible (Table 4-1). (The term *single-gene disorder* refers to the involvement of a single genetic locus; for any given locus a pair of genes (alleles) work together, except for

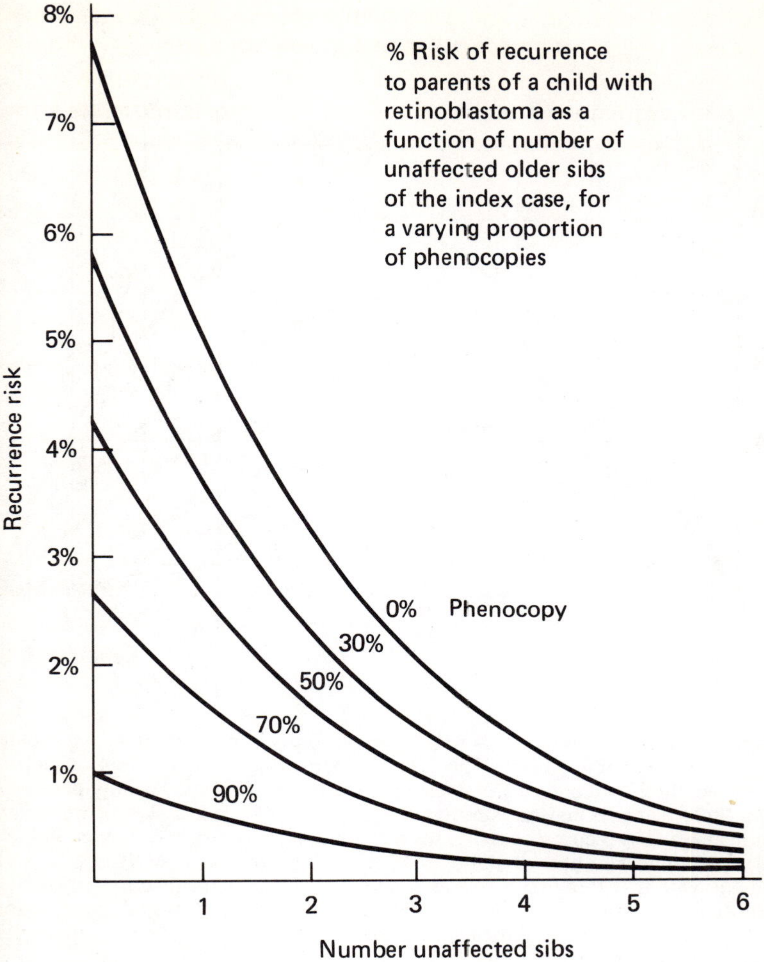

Figure 4–1 Sibling recurrence risks for retinoblastoma (autosomal dominant trait with frequent phenocopies). For any percentage of phenocopies, increasing numbers of normal siblings decrease the likelihood of recurrence among future siblings. The parents are presumed to be unaffected after reliable negative examination. This graph emphasizes the importance of combined genetic and empiric data for assessing recurrence risks. (Courtesy of R. Nussbaum and J. Puck.) Retinoblastoma may or may not be an autosomal dominant trait (i.e., there are phenocopies) (46). Thus the need for this graph. It may be used for other autosomal dominant traits with similar penetrance and phenocopy concerns.

the case of the sex chromosomes in a male. Thus, "single gene" disorders or traits really refer to the interplay of a pair of genes or alleles.) Ordinarily, both genes at a given locus are of the normal, or wild-type, variety; they both represent the normal or wild-type allele (see glossary, p. 63). When a change occurs in a gene, we speak of a mutation; the result is a mutant gene, occurring as an

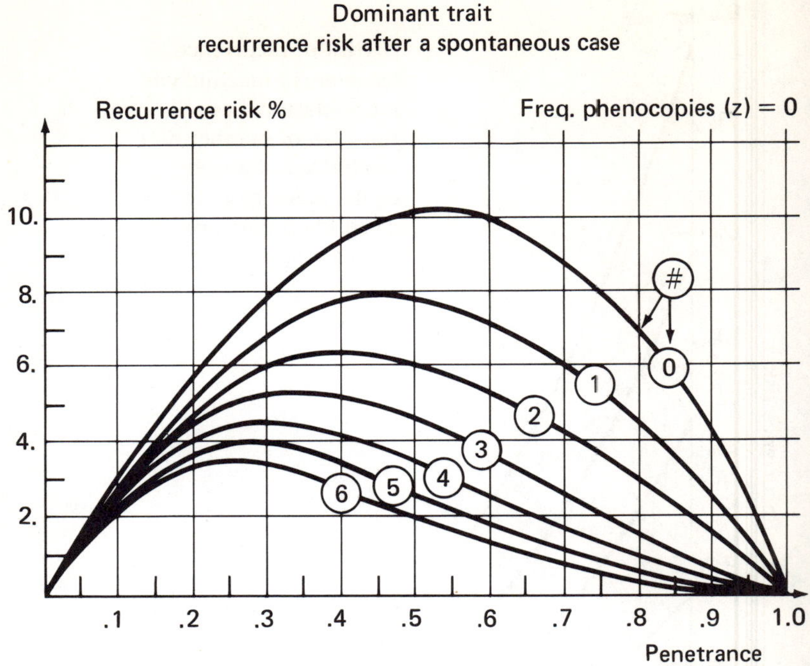

Figure 4–2 Sibling recurrence risk for an autosomal dominant trait as a function of penetrance. For any level of penetrance, increasing numbers of normal siblings decrease the likelihood of recurrence among future siblings. For low penetrance (e.g., 0.5) even six normal siblings leave a 2% recurrence risk and no siblings indicates a 10% recurrence risk. This formulation presumes no phenocopies (Z = 0). # = number of normal siblings to proband. (Courtesy of R. Nussbaum and J. Puck.) Note that this graph relates to autosomal dominant disorders in general, and that the parents are presumed to have negative histories and examinations. Note that penetrance is indicated as the proportion of persons with the mutant gene who express it.

allele to the wild-type gene at that locus. The mutant gene's expression (phenotype) is determined to be different on biochemical or functional (clinical) grounds. If the mutant allele (or its expression) is overshadowed or masked by the wild-type allele, the mutant is said to be recessive. That is, the heterozygote, with one mutant and one wild-type gene, does not clinically manifest the mutation. In contrast, dominant mutations overshadow or mask the wild-type allele, so that they are clinically manifested in the heterozygote. Although there are intermediate cases where both genes are equally expressed as codominants (e.g., ABO blood-group alleles), and ultimately the expression of any mutant gene can be defined biochemically or immunologically, the concepts of dominance and recessiveness have great clinical import.

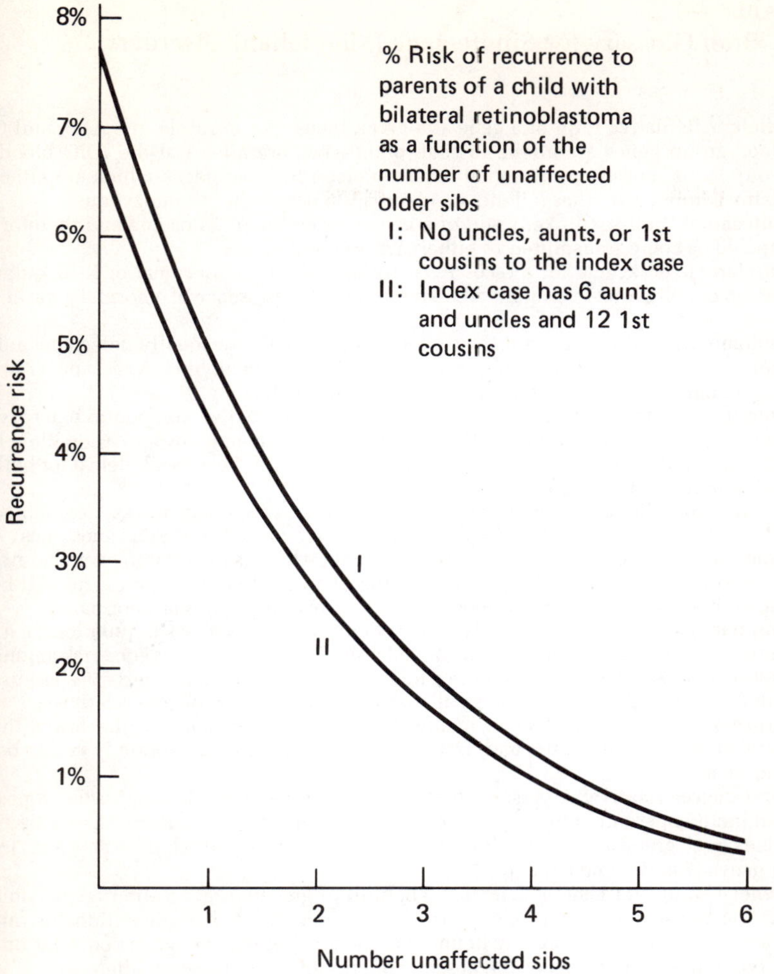

Figure 4–3 Sibling recurrence risks for bilateral retinoblastoma (autosomal dominant trait with incomplete penetrance). While recurrence risks for future siblings to the proband decrease rapidly as a function of unaffected siblings (first-degree relatives), they are changed little by negative findings in six aunts and uncles (second-degree relatives) and twelve first-cousins (third-degree relatives). (Courtesy of R. Nussbaum and J. Puck.) Bilateral retinoblastoma represents an autosomal dominant mutation unless a deletion of chromosome #13 is demonstrated (46); as implied here, penetrance may be less than 100%. This graph may be used for other autosomal dominant disorders with similar penetrance concerns.

If a characteristic is clinically apparent in the heterozygote (dominant) and the affected person can procreate, the disorder will be transmitted with a 50% probability even if the other parent is normal. If a characteristic is not expressed in the presence of a wild-type gene but is expressed if both alleles at that locus are

TABLE 4-1
A Brief Glossary for Single-gene (Mendelian) Disorders

Allele: Alternative form of a gene at a given locus. For example, the A, B, and O blood-group genes are allelic to each other—i.e., are alleles at the ABO blood-group locus. For each autosomal locus a person has two genes which are either distinct from each other (allelic) or identical to each other (homozygous).
Autosomal: Non-sex-linked; said of a gene carried on an autosome and therefore capable of being transmitted to either sex from either sex.
Carrier: Heterozygote for a recessive disorder, whether autosomal or X-linked; a person or cell with one mutant recessive gene in the presence of a normal gene at a given locus.
Codominant: Alleles are said to be codominant if both are equally detectable and their effects are additive; the genes (alleles) for *hemoglobins* A, S, and C are codominant, as are those for alpha-1-antitrypsin deficiency.
Compound heterozygote: An individual, or cell is said to be a compound heterozygote when he possesses each of the two alleles for a given locus and each allele is mutant and different from the other; e.g., a person with the genes (alleles) for both hemoglobin S and hemoglobin C.
DNA: Deoxyribonucleic acid, the nucleic acid genes are made of; DNA combined with specific proteins constitutes chromatin, which makes up the chromosomes.
Dominant: A mutant gene is said to be dominant when its expression overshadows or masks the normal wild-type allele. A disorder due to a dominant gene will be transmitted with a 50% likelihood if the other parent's genes are normal.
Empiric risk: Risk estimates derived by actually sampling at-risk populations; e.g., in the presence of an otherwise negative family history, the recurrence risk among first-degree relatives of a child with a neural tube defect is said to be 3–5% because among 100 similar families 3–5 will show an affected first-degree relative.
Expression: The sum of gene manifestations for a given locus; the fewer the number and severity of the manifestations, the milder the expression is said to be, and vice-versa.
First-degree relative: A person's first-degree relatives share 50% of his/her genes, and include parents, siblings, and offspring. Second-degree relatives share 25% of genes (e.g., grandparents, uncles, nieces, etc.). Third-degree relatives share 12.5% of genes (e.g., first cousins, etc.).
Gene: The unit of genetic function. The unit of genetic organization responsible for the constancy of a property or phenomenon from one generation, or one person, to another. Ultimately defined in chemical terms (one gene codes for one polypeptide) or in terms of the absence of crossing over between alleles.
Genetic heterogeneity: When there is more than one genetic mechanism to account for a disorder—e.g., two or more mutant alleles at a given locus (cystinuria), or more than one locus (hemophilias A and B, etc.). Determination of heterogeneity often leads to the distinction of two separate disorders (e.g., tyrosinase-positive versus tyrosinase-negative albinism).
Genotype: The actual genetic makeup of an individual or cell, either with regard to a general, but defined, set of considerations (e.g., lipid metabolism, etc.), or with regard to a specific locus (e.g., blood groups). This concept underlies the notion that more than one combination of genes may correspond to a given phenotype.
Hemizygous: A person or cell is said to be hemizygous for a given genetic locus when there is only one allele present. The second allele is absent either because only one of the two chromosomes or its segments is present—e.g., in males where there is only one X-chromosome and in chromosomal rearrangements resulting in a deletion (e.g., 18q−).
Heterozygote manifestation: An individual heterozygous for a recessive mutation ordinarily does not clinically manifest the mutant gene; if the recessive mutation is considered to be clinically though subtly obvious, we refer to heterozygote manifestation.
Homozygous: A person or cell is said to be homozygous for a given genetic locus when both alleles are identical.

Incest: Consanguinity involving a mating between first-degree relatives (brother–sister, mother–son, or father–daughter) or between second-degree relatives.
Locus: The actual location of a gene along the length of a chromosome. This term applies whether or not the specific chromosome involved can be designated.
Lyon principle: The notion that in any given cell only one X-chromosome is genetically active and that therefore any others present have been inactivated. The inactivating process occurs within days after conception, is random from one primitive cell to another and is *constant* for that cell and all its daughter cells.
Meiosis: The final sequence of cell divisions whereby germ cells are formed. In the penultimate division (meiosis I) the homologs of a pair are separated from each other (reduction division). In the final division (meiosis II) the chromatids for each homolog separate from each other. The resulting egg or sperm has the haploid number of chromosomes (23). Crossing over occurs in meiosis I (compare mitosis).
Mendelian: Refers to genes and resultant traits or disorders being transmitted according to specific patterns as elucidated by Gregor Mendel; Mendalian disorders are also referred to by the term "single-gene" disorders.
Mitosis: The process of cell division in somatic cells. Each daughter cell winds up with the same number of chromosomes as the parent cell (46) (compare meiosis).
Mutation: A change in a gene's expression and chemical makeup.
Mutation rate: The rate at which a given locus is known to undergo mutation; determined empirically.
Pedigree: The stylized, graphic way of portraying a family history; family tree.
Penetrance: Whether or not a dominant mutation is expressed at all. If expressed, no matter how slightly, the gene is penetrant, and if not expressed it is nonpenetrant.
Phenocopy: A nongenetic trait or disorder which mimics a genetic one.
Phenotype: The observable or measurable expression of a gene or genes.
Proband: The affected individual whose disorder brings a family to genetic evaluation and counseling.
Recessive: A mutant gene is said to be recessive when its expression is masked or overshadowed by the normal, wild-type allele. Clinically, a recessive gene is ordinarily apparent only when combined with another mutant allele at the same locus, or in the hemizygous state.
RNA: Ribonucleic acid, the nucleic acid which is involved in various types of cell metabolism and especially the conversion of potential information in genes to actual gene products and gene expression.
Sex-linked: A gene carried on a sex chromosome is sex-linked. Most often this term implies X-linkage, since there is only a very small number of specifiable genes on the Y-chromosome.
Single-gene: Mendelian.
Wild-type: The nonmutated allele at a gene locus; usually the commonest allele. This term can be helpful since it implies that a mutant gene is not automatically considered to be pathologic (i.e., not normal), rather merely distinct from the wild-type.
X-linked: A gene carried on the X-chromosome is said to be X-linked.

mutant (recessive), then recurrence among siblings to the affected person depends on both parents transmitting a mutant gene and the likelihood of this is 1 in 4. In addition, both dominant and recessive disorders are compatible with an apparently negative family history (Table 4-2).

Patterns of Inheritance

We can also characterize mutations as to whether the locus is on a sex chromosome (sex-linked) or an autosome (autosomal). Until

TABLE 4-2
Reasons Why a Genetic Disorder May Be Associated with a Negative Family History

1. Incomplete history
2. Nonpaternity
3. New mutation
4. Variable expression, variable penetrance
5. Recessive or polygenic traits

very recently only one gene was specified as having a Y-chromosome locus, so that "sex-linked" used to be synonymous with "X-linked." Now that Y-linked genes are being described (2), however, we have to be more particular about distinguishing "X-linked" from "sex-linked." Dominant and recessive genes are known for both autosomes and the X-chromosome. Thus, there are autosomal recessive, autosomal dominant, X-linked recessive, and X-linked dominant alleles and disorders.

An affected individual with an *autosomal dominant disorder* (Fig. 4-4 and Table 4-3) has a 1 in 2 chance of passing on the mutant gene to each offspring regardless of the offspring's sex; male-to-male transmission is an important feature of autosomal dominant characters. This 50% risk is present even if the spouse is genetically normal. Artificial insemination by donor can avoid this transmission risk for an affected male, but obviously not for an affected female. Other characteristic features of autosomal dominant traits include multiple-generation involvement (Fig. 4-3), spontaneous mutations, and the possibility of marked variation of expression. From a clinical point of view, however, the critical thing is that whoever has the mutant gene has a 50% risk for passing it on to each offspring, even if the other parent is normal. Among the disorders that fall into this category are achondroplasia, Huntington chorea, neurofibromatosis, and Von Willebrand disease.

Autosomal recessive disorders (Fig. 4-5 and Table 4-3) are manifested when both alleles at a given locus are mutant. If both mutants are "exactly the same" the individual is a homozygote (e.g., sickle-cell anemia), while if they represent two different mutations the person is a compound heterozygote (e.g., hemoglobin SC disease). If the parents of an affected child are genetic relatives (consanguineous), homozygosity is presumed. In either case a mutant gene must be contributed by each parent. Both sexes are affected and recurrence risks in each subsequent pregnancy are: 1 in 4 for a similarly affected child, 2 in 4 for a carrier (heterozygote) child, and 1 in 4 for a homozygous normal child. Any previously born, unaffected siblings of the affected person would each have a 2 in 3 chance of being a carrier and a 1 in 3 chance of being homozygous normal. In contrast with dominant disorders recurrence risks are high (25%) only for siblings of the affected person, except for three special sets of circumstances—subsequent consanguinity, marriage into another family with the same disorder, or a relatively high carrier rate (i.e., 1 in 10 for sickle-cell anemia). If the affected person procreates, all offspring are carriers (heterozygotes); the likelihood of having an affected offspring would then be one-half the carrier rate in the general population. Thus, if a woman with cystic fibrosis (CF) conceives, and the father is a Caucasian with a negative family history (carrier likelihood then is about 1 in 20), the chance of having an affected child is 1 in 40: (probability that mother will transmit CF mutation = 1) × (probability that father is a carrier for CF = 1/20) × (probability of father transmitting CF gene if it is present = 1/2) = 1/40. For normal siblings of the affected woman the chance is

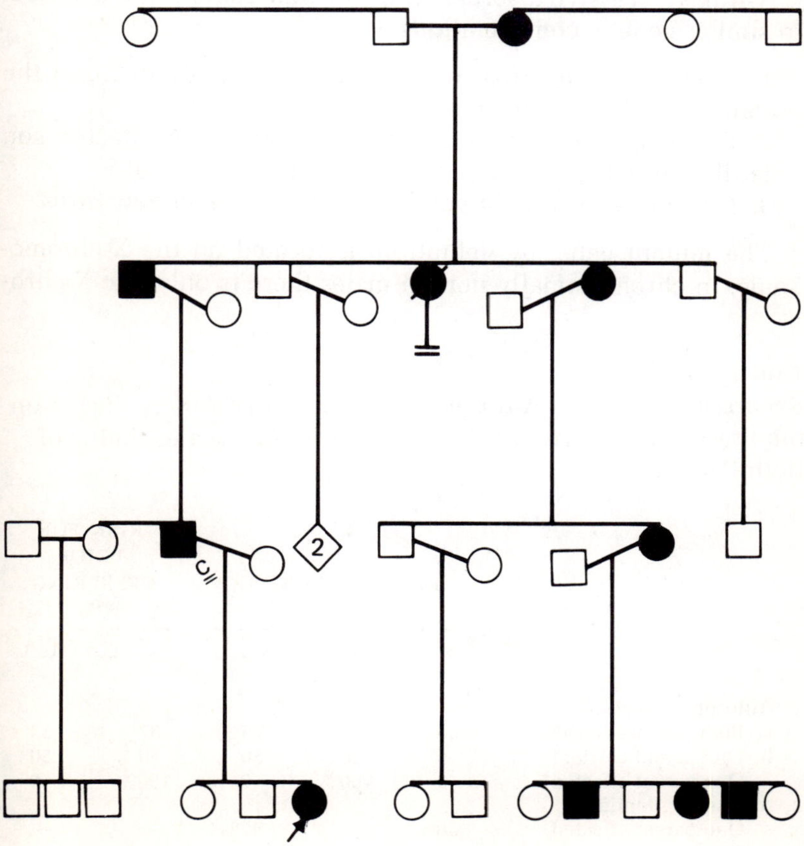

Figure 4–4 Characteristic autosomal dominant pedigree. Note almost equal male and female involvement in multiple generations and male-to-male transmission. For symbol meanings see Fig. 9–1, p. 174.

even less: (probability that sibling is a CF carrier = 2/3) × (probability of sibling transmitting CF gene, if it's present = 1/2) × (probability that spouse from general population is also a carrier = 1/20) × (probability that spouse will transmit CF gene if it is present = 1/2) = 1/120. The major points about autosomal recessive disorders are thus:

1. Both parents must contribute the mutant gene.
2. The likelihood of recurrence in subsequent generations is relatively small unless certain unusual circumstances prevail.

Artificial insemination by donor is an important and simple means of markedly decreasing the recurrence risk among siblings to a child affected with an autosomal recessive disorder. This will be discussed at length in Chapter 11. Prenatal diagnosis methods, especially amniocentesis, can also be utilized for increasing numbers of these disorders (see Chapter 8).

X-linked recessive disorders (Fig. 4-6 and Table 4-4) are subject to similar genetic considerations:

1. The carrier (heterozygote) does not ordinarily manifest the mutant gene (i.e., it is recessive).
2. She will have a 1 in 2 chance of having an affected son regardless of whether her spouse is genetically normal.
3. New mutations are important in accounting for new cases.

The mutant gene, by definition, is located on the X-chromosome. In chromosomally normal males there is only one X-chro-

TABLE 4-3
Recurrence Risks for Autosomally Inherited Disorders, Based on the Presence of an Affected Individual or the Carrier Status of Both Parents

	RISK FOR FUTURE OFFSPRING, %			RISK FOR PRIOR OFFSPRING, %	
	A	*C*	*NCA*	*C*	*NCA*
1. Autosomal recessive					
a. Both parents carriers	25	50	25	67	33
b. One parent carrier*	0	50	50	50	50
c. One parent affected*	0	100%	0	100%	0
2. Autosomal dominant					
a. One parent affected	50%	—	50%	—	—
b. Neither parent affected	μ	—	$1 - \mu$	—	—

Note: recurrence risks depend on the type of inheritance pattern and whether one or both parents have the mutant gene in question. A = affected, C = carrier, NCA = neither carrier or affected, μ = mutation rate, * = other parent presumed not to be a carrier.

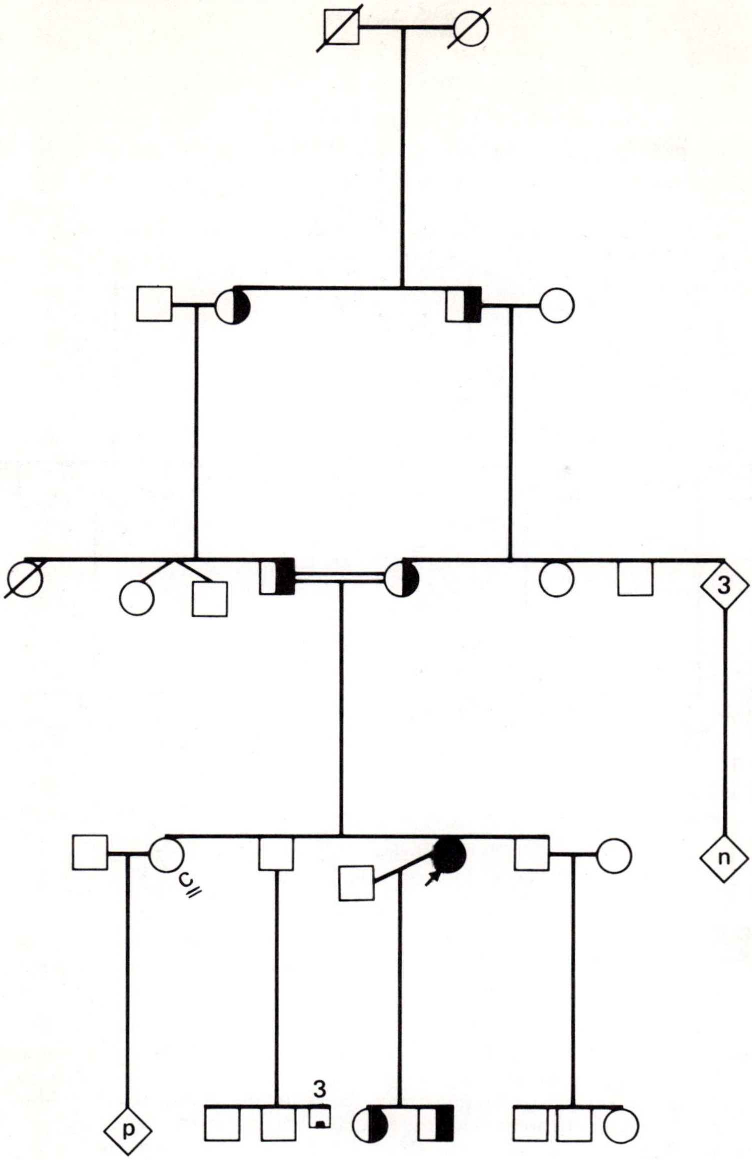

Figure 4–5 Characteristic autosomal recessive pedigree. Note the consanguinity and that both parents and offspring of the proband are obligate heterozygotes (carriers). For symbol meanings see Fig. 9–1, p. 174.

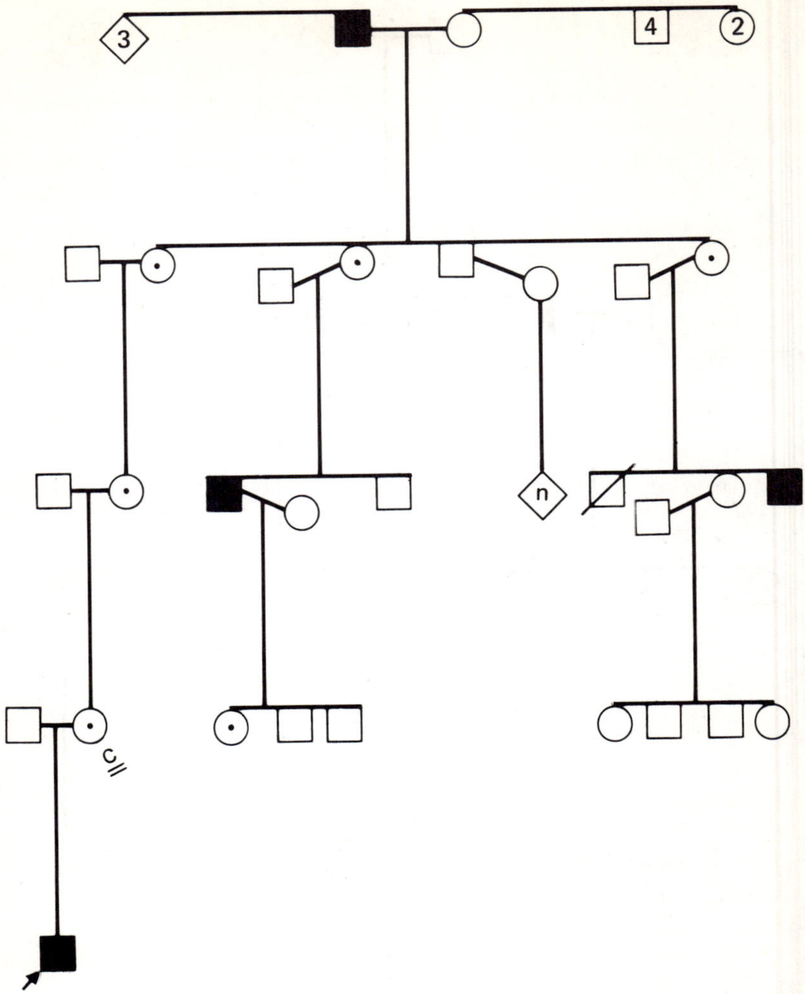

Figure 4–6 Characteristic X-linked recessive pedigree. Note that only males are affected and that they are related through unaffected females. As indicated on the reader's far left, several generations may elapse before an affected individual is born in the lineage of carrier females. All daughters of an affected male must be carriers and none of his sons can be affected. For symbol meanings see Fig. 9-1, p. 174.

mosome, so that even a recessive mutation is expressed if present (hemizygosity). If the affected male inherited the mutant gene from his mother (i.e., she is a carrier), then each subsequent brother to the proband will have a 1 in 2 chance of having the disorder and each sister a 1 in 2 chance of being a carrier. A brother without signs of the disorder does not have the mutant gene and thus cannot pass it on. For sisters who are not carriers there is no increased risk. One of the recurring dilemmas in clinical genetics, however, is knowing whether the mother and sisters of a male with an X-linked recessive disorder, when the family history is apparently negative, are actually carriers for the disorders. The reason for the dilemma is that the mutation could have occurred in the affected male, and was therefore not actually inherited, or in his mother (or her mother), making her a carrier. Only in the latter circumstances would sisters or maternal aunts actually have a risk for being carriers. If there are straightforward biochemical or immunologic methods of identifying carriers, the dilemma is minimal (as in the Lesch-Nyhan syndrome or Hunter syndrome). In many instances, however, there is only partially reliable carrier detection (e.g., Duchenne muscular dystrophy, hemophilia A) or none at all (e.g., anhydrotic ectodermal dysplasia, nonspecific X-linked mental retardation or Renpenning syndrome). If detection is partial or nonexistent, various types of empiric data (i.e., mutation rates and number of unaffected family members) must be used to establish carrier and recurrence risks. The finer details of these considerations will be dealt with in specific examples below.

If a male affected with an X-linked recessive disorder procreates, he can transmit the mutant gene to *none* of his sons, but he will transmit it to *all* of his daughters. This point cannot be overemphasized: All daughters of a man with an X-linked recessive disorder must be carriers. Every daughter of a man with hemophilia A will be a carrier and each of her sons will have a 50% risk for the disorder. A pedigree of a family with an X-linked recessive disorder that does not represent a new mutation usually shows that 50% of all at-risk males have been affected, all of them related through nonaffected carrier females.

A special consideration regarding manifestations of such a disorder in carrier females relates to the Lyon principle: It is

TABLE 4-4

Recurrence Risks for X-Linked Inherited Disorders Based on the Presence of an Affected Individual or the Carrier Status of the Mother

	MALE OFFSPRING				FEMALE OFFSPRING			
	A	*NA*	*C*	*NC*	*A*	*NA*	*C*	*NC*
1. X-linked recessive								
a. Father affected*	0	100%	—	—	—	—	100%	0
b. Mother carrier*	50%	50%	—	—	—	—	50%	50%
2. X-linked dominant								
a. Father affected	0	100%	—	—	100%	0	—	—
b. Mother affected	50%	50%	—	—	50%	50%	—	—

Note: With X-linked traits the recurrence risk is a function of both the sex of the parent with the mutant gene and the sex of the offspring. A = affected, NA = not affected, C = carrier, NC = not a carrier, * = other parent presumed not to have the mutant recessive gene.

possible with any heterozygous female that by chance in all, or almost all, of her cells the X-chromosome which carried the normal (wild-type) allele has been inactivated. In this case the woman might well manifest all the features seen in the hemizygous male. This can be confusing since the inheritance pattern will seem to be something other than X-linked recessive. For a detailed account of the Lyon principle the reader is referred elsewhere (3). For all clinicians, however, it is important to realize that this phenomenon occurs often enough to be clinically significant. Not only is it relevant to genetic counseling, but it may also influence immediate management considerations; for example, a female carrier of hemophilia A may require Factor VIII replacement before major surgery (4).

On the other hand, it is the unexpected wrinkles like this that can make it so frustrating for the nongeneticist to deal with genetic problems. There always seems to be an exception or unanticipated consideration. In this regard I would emphasize that a clinician does not need to know all the regular patterns and ordinary expectations, as well as the anticipated exceptions. Rather, it is merely necessary to acknowledge that there *are* exceptions and that it is not always easy to recognize exceptional situations when they are present.

In addition to Lyonization, other explanations may account for the manifestation of an X-linked recessive disorder in a female, although all but the first of those that follow would entail a second pathologic process:

1. An affected male had a daughter by a carrier female, i.e., the daughter was homozygous.
2. The woman's chromosome constitution was 45,X.
3. The woman had a 46,XY chromosome constitution (e.g., testicular feminization).
4. Genetic heterogeneity could be involved, the affected female having an autosomal recessive disorder that was being confused with the X-linked recessive one. Such overlap did occur as the two diseases involving diffuse angiokeratomata (Fabry disease and fucosidosis) were being sorted out (5).

X-linked dominant disorders (Fig. 4-7 and Table 4-4) are relatively uncommon. They are characterized by multiple-generation involvement, a ratio of affected females to males of 2 to 1, absence of male-to-male transmission, and greater severity in males than females. In contrast to X-linked recessive disorders, there are no unaffected carriers; heterozygotes are expected to manifest the disease. However, the X-linked restrictions on the transmissability of the gene apply. Each daughter and each son of an affected woman would have a 1 in 2 chance of having the disorder. None of the sons of an affected male would be affected, but all of his daughters would have the disorder. That is, he could not transmit

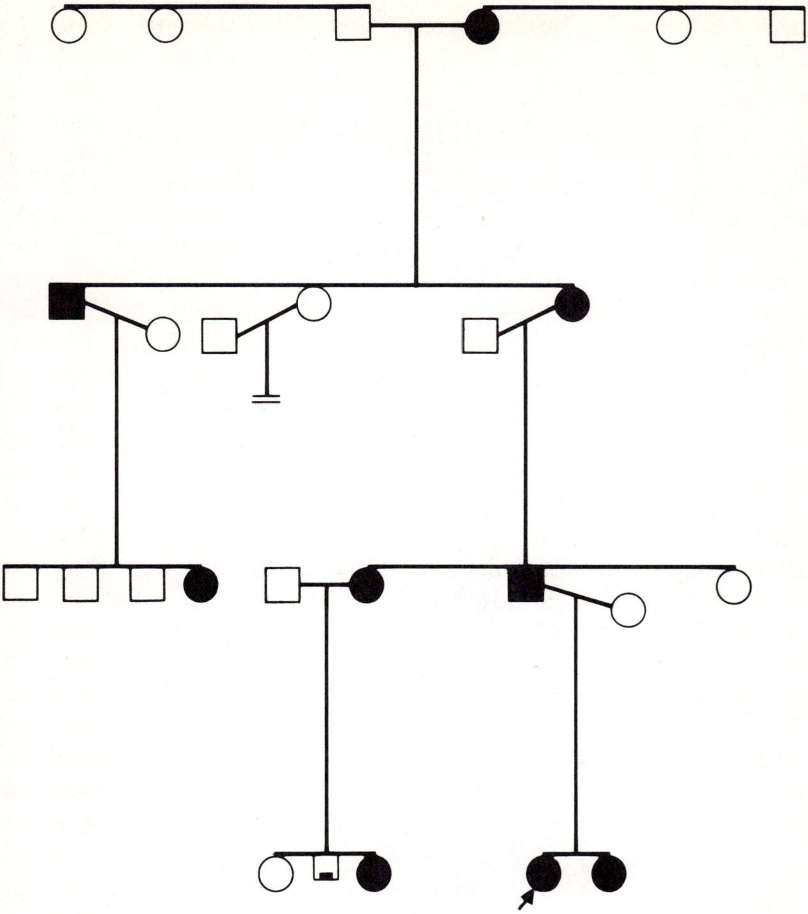

Figure 4–7 Characteristic X-linked dominant pedigree. Note that multiple generations are involved and the number of affected females exceeds the number of affected males by at least 2:1. There is no male-to-male transmission and all daughters of an affected male must be affected. For symbol meanings see Fig. 9–1, p. 174.

an X-chromosome to his sons but must transmit it to his daughters (presuming they are karyotypically normal). Phosphate diabetes has this inheritance pattern. Albright hereditary osteodystrophy (pseudohypoparathyroidism) has long been thought to have an X-linked dominant pattern (6), but this is not entirely certain (7).

Because X-linked disorders are not uncommon (e.g., hemophilia and Duchenne muscular dystrophy) and because there is differential transmission of the mutant gene to either sex, prenatal diagnosis has special, even controversial, importance for these disorders. This will be considered further in the examples to follow and in Chapter 8.

Y-linked traits exist, including hairy pinnae (8) and the Y-linked antigen (2), but no specific disorders are known to be Y-linked.

Mutation, Variability, and Heterogeneity

At the end of this chapter, some selected disorders will be discussed as clinical examples of Mendelian principles. First, however, it is important to discuss mutation. variability, and heterogeneity. Consider, for example, the question of recurrence among siblings of a child with an apparently heritable disorder born to presumably normal parents with a negative family history and no known consanguinity. There are several possible ways to explain heritability in this context (Table 4-2): inaccurate or incomplete family history, nonpaternity, decreased expression to subclinical levels in either parent, new mutatation, genetic heterogeneity, phenocopy, and recessive or polygenic traits. Each of these possiblities could account for any case, but some are more likely to apply to certain diseases. For example, a *new mutation* is most likely with achondroplasia (because that mutation is virtually always expressed) or tuberous sclerosis (because it is usually a genetic lethal not allowing for procreation). For at least certain dominant disorders, advanced paternal age may increase the likelihood of a new mutation (9). *Decreased expression* is very likely with hereditary spherocytosis, the Waardenburg syndrome, or Conradi disease. With isolated retinitis pigmentosa, which can be an autosomal recessive, autosomal dominant, or X-linked recessive trait, *genetic heterogeneity* is the important consideration. Unilateral retinoblastoma may represent a *phenocopy* (Fig. 4-1).

A characteristic of almost all dominant disorders is their variable expression from one person to another, even within the same family. At one extreme are cases with no apparent expression of the mutation, even though the person's father, siblings and offspring are affected and nonpaternity has been excluded. This is referred to as *nonpenetrance.* For any given dominant mutation a characteristic proportion of heterozygotes will fall in this category; this proportion defines the penetrance of the mutation. Thus, if only half of those with the mutant gene manifest the disorder, the penetrance is said to be 50%. The underlying mechanisms that account for decreased penetrance are not at all understood, and thus the degree of penetrance characteristic of any given mutation must be determined empirically. This determination can be critically important, and it injects caution into many genetic counseling considerations.

The general point, however, is that with significantly decreased penetrance we must talk both about the 50% risk for passing on the dominant mutant gene and the lesser risk of passing on the disorder itself. Conversely, lack of expression does not preclude transmission of the disorder. An apparently normal person in a sibship with a mutation of notoriously low penetrance (e.g., brachydactyly, type D) would still have a substantial risk of transmitting both the mutant gene and the disorder.

In a pedigree chart the individuals in whom the gene is *nonpenetrant* are not marked as expressing the gene; that is, they have *no* phenotype stigmata despite their genotype. In contrast, those who are designated as expressing the gene only partially represent *decreased expression.* Persons with decreased expression are identified by their own mild features, while those with nonpenetrance are recognized from other data, including affected offspring, siblings, parents, etc.

The concepts of variable expression and variable penetrance, as used above, are usually restricted to dominant mutations and disorders. Recessive disorders are ordinarily much less variable from one person to another in a given family, and even from one family to another. If there were consistent differences between families, other explanations would have to be considered, including genetic heterogeneity (i.e., different loci involved, as with tyrosinase-positive and tyrosinase-negative albinism) and compound heterozygotes (i.e., two or more mutant alleles at a given locus, as with cystinuria types I, II, and III alleles). If an individual is homozygous or hemizygous for a given mutant allele, therefore, the full-blown clinical disorder is expected even though the details may be slightly different from one person to another; and the idea of nonpenetrance does not apply. There are very few exceptions to this, but some of them are important—phenylketonuria (PKU), cystic fibrosis, and alpha-1-antitrypsin deficiency (AATD). Some persons with PKU may be of normal intelligence and yet have all the characteristic biochemical aberrations (see Chapter 6). Some persons with cystic fibrosis may have neither gastrointestinal nor pulmonary problems through early adulthood, only to develop a picture of "chronic bronchitis" later on. Some homozygotes for AATD may show severe hepatic dysfunction in infancy or childhood, while a similarly endowed sibling may show nothing or later develop chronic obstructive pulmonary disease and emphysema.

Some Specific Disorders

The following sketches of Mendelian disorders (see Table 4-5) are arranged first according to mode of inheritance, and then, roughly, according to timing and manner of presentation, beginning with congenital or infantile disorders. The intention is to indicate that while there are many types of heritable disorders, there is also a certain consistency among them.

TABLE 4-5
Some Mendelian Disorders That Are Relatively Common or Well Known

Adrenogenital syndrome	Hyperlipidemia
Alpha-1-antitrypsin deficiency	Isoniazid acetylation defect
Brachydactyly, type E	Neurofibromatosis
Color blindness	Phenylketonuria
Cystic fibrosis	Polycystic kidney disease of adult onset
Deafness, severe congenital	Polysyndactyly
Duchenne muscular dystrophy	Retinoblastoma
Galactosemia	Sickle-cell anemia
Glucose-6-phosphate dehydrogenase deficiency	Stickler syndrome
Hemophilia A and B	Tay-Sachs disease
Hereditary spherocytosis	Thalassemia
Huntington chorea	Von Willebrand disease

Autosomal Dominant Disorders

Conradi disease has diffuse manifestations that are usually obvious at birth (10). The cardinal features are an asymmetric shortening of the limbs (proximal, middle, and distal segments), limitation of motion at major joints, widespread erythematous scaling of the skin, alopecia, trichorhexis, and a rounded facies somewhat suggestive of the Down syndrome. Radiographically, there are punctate calcifications of the epiphyses. The first task is to distinguish this disorder from the similar autosomal recessive trait with the same name, but with a different radiological appearance and a more severe clinical course. Once the heterogeneity is clarified, then it must be decided whether the defect is a new or an inherited character, if possible. If either parent was similarly affected, then inheritance is obvious and the recurrence risk is 1 in 2. If neither parent is evidently affected and there are no siblings to the proband, mild expression or nonpenetrance could not be excluded. For the proband's future siblings, the risk of having the mutant gene *could* be as high as 50%, and a certain proportion would also have clinical problems if this is the case. On the low side, the risk for siblings is essentially that of the general population if the defect is a new mutation. The birth of a second affected child would verify nonpenetrance. In either case, the recurrence risk for each of the proband's offspring is 1 in 2.

The *Holt-Oram syndrome* is characterized by congenital heart disease (usually an atrial septal defect) and reduction defects of the thumbs or, alternatively, triphalangeal thumbs (11). It is usually obvious at birth, though at times the defects of the heart and extremities can be subtle. The main point is that such thumb abnormalities should lead to a very careful cardiac examination, and if a lesion is found, all first-degree relatives should be examined to determine, if possible, what the risk for future siblings would be.

Polysyndactyly is manifested only in the limbs. In the full-blown syndrome there is postaxial polydactyly of the hands and preaxial polydactyly of the feet (great toes); broad and stubby thumbs; syndactyly of the hands involving digits III, IV, and V; and syndactyly of the feet involving digits I, II, and III (12). Diagnosis provides reassurance that there are no associated visceral anomalies. Penetrance is usually complete, but expression can be very variable from one person to another, even in the same family; one person may only show the postaxial polydactyly, another only broad great toes and foot syndactyly, etc. New mutations are not uncommon.

The *Marfan syndrome* is an autosomal dominant trait that usually becomes apparent in childhood or early adult life (13). Classically, its main features are a tall, thin body, arachnodactyly, hyperextensible joints, scoliosis, pectus deformities, dislocated lenses, retinal tears, mitral-valve deformities, and dissection of the aortic arch. Diagnosis is usually made on the basis of at least four or five of these features in one person, or even fewer if there is a positive family history. Early diagnosis to minimize the occurrence or consequences of complications (e.g., retinal tears, aortic dissection, scoliosis) is important. The burden of the disorder is heavy, and affected persons might well elect to limit family size if informed of the 50% recurrence risk among offspring, especially considering the possibility of death in early adulthood from the cardiac complications. Distinguishing nonspecific phenocopies from mild cases of the Marfan syndrome can be frustratingly difficult. It is also important to differentiate the syndrome from other heritable disorders, especially homocystinuria and the Ehlers-Danlos syndrome.

The *Stickler syndrome* is thought to be quite common (14). Its main features usually become manifest in childhood and early adult life: high myopia, deafness, and arthropathies involving the major joints. It may also present at birth as the Pierre-Robin anomalad (severe micrognathia, with secondarily recessed, floppy tongue and cleft palate). All families of infants with the Pierre-Robin anomalad should be investigated, at least by history-taking, for the presence of this syndrome (14). Expression can be very variable, but the burden is very great in some cases.

Von Recklinhausen neurofibromatosis (NFT) represents a defect in cells derived from the neural crest (15). It is relatively common, with a frequency as high as 1/3200. The mutation rate is thought to be high, as many as 50% of the cases having a previously negative family history. However, perhaps only 5% of mild cases are identified (16). The primary features include cutaneous cafe-au-lait spots (hyperpigmented macules) and neurofibromas in both the peripheral and central nervous systems. Severe complications are frequent. They include congenital tibial pseudarthrosis, intracranial and orbital tumors (childhood), severe kyphoscoliosis (childhood), severe cosmetic disfigurement from increasing numbers and/or size of neurofibromas, and in some cases malignant degeneration of neurofibromas. This disorder in some persons is subtle and in others devastating. Among the worst tragedies is that a very mildly affected person may beget a severely affected child, often without having been aware of the 50% risk of passing on the trait in one form or another. Its range of expression is impressive and should be more widely known. If 95% of mild cases are usually missed and yet each has a 50% risk of transmission and a 25% chance of a moderately or severely affected offspring, much disease could be prevented by keener diagnosis and increased availability of genetic counseling. The extreme variability of NFT exemplifies the importance of this phenomenon for autosomal dominant disorders in general.

Pseudoxanthoma elasticum (PXE) is usually inherited as an autosomal dominant trait and often presents in early to middle adult life (17). It has protean manifestations, including hyperelasticity of the skin; angioid streaks and hemorrhage of the retina; gastrointestinal hemorrhage; peripheral vascular occlusive disease; myocardial infarction; and typical skin changes on the nape of the neck, axillae, and antecubital and popliteal spaces. These skin lesions have the appearance of plucked chicken skin or rough-grained leather. When the findings are characteristic and two or more generations are involved, the genetic counseling is straightforward. However, several considerations confound counseling when only one person is affected: 1. As with other dominant traits, expression can be very variable and spontaneous mutation is possible. 2. There is considerable genetic heterogeneity, with at least two dominant and two autosomal recessive forms recognized (18) and no unequivocal way of distinguishing them in any given case on clinical or laboratory grounds. 3. Environmental factors may be critical in determining the expression of the disorder. Thus, in the not infrequent situation of only one affected family member, genetic counseling is not at all clear. For the dominant form there would be a 50% recurrence risk among offspring, but for the recessive form there is no risk among offspring unless marriage to a genetic relative occurs. In any case the situation is complicated and may remain unresolved even after elaborate evaluations. This limitation is important in counseling for many other heritable disorders as well.

Polycystic kidney disease of the adult onset variety (PKD) is an autosomal dominant disorder with two primary manifestations: multiple cystic lesions of the kidneys and liver and intracranial berry aneurysms (19). Either may ultimately cause death. The renal and hepatic cysts may go undetected for many years, at times being discovered only through family studies or coincidentally at autopsy or clinical examination for other reasons. Hypertension, hematuria, or renal failure are frequently the presenting problems, usually after age 20. Often a person has long since completed his or her family before the diagnosis is made, and the genetic counseling effort then centers on identifying which offspring are affected so that optimal management can be initiated and the prospect of prevention afforded future generations. Since renal transplantation is an important therapeutic method, the documentation of unaffected family members as potential donors is part of the family evaluation. When the correct, minimal diagnostic criteria are used (urinalysis and intravenous pyelogram or renal ultrasound examination) and the individual's age is taken into account, penetrance is essentially complete. This allows fairly realistic genetic counseling for all family members. In children and infants it is of course important to distinguish this disease from others involving renal cysts, and especially infantile multicystic disease (an autosomal recessive disorder) and renal dysplasia with cysts (virtually always a sporadic developmental disorder).

Huntington chorea becomes manifest in adult life, usually after age 30 (20). Thus, families are usually begun and often completed before this devastating disorder becomes apparent. It is characterized by deterioration of intellectual functions (dementia) and involuntary movements of the limbs, face, and trunk (chorea). Once manifest, there is steady deterioration, leading to death after some years. There is neither effective treatment nor reliable means of detecting the presence of the mutant gene before it becomes clinically apparent. And this is a critical point for genetic counseling: most often those who request counseling are apparently normal offspring of patients with the disease. Their own chance for developing the disorder is basically 1 in 2, though in reality it may be less than that because empirically we can determine a decreasing likelihood of ever showing the disorder as a function of age. For example, a nonsymptomatic 50-year-old son of an affected parent has a lower empiric risk than his nonsymptomatic 25-year-old brother. For a 20-year-old son of an affected parent the risk is essentially 1 in 2, so that his risk of having affected children is half that ($1/2 \times 1/2 = 1/4$). If he never develops the disorder, his children are not at risk. Having to acknowledge the 25% risk for affected offspring while perfectly healthy in the meantime can be very frustrating for these at-risk persons. So genetic counseling affords more than just an appraisal of risks; it is also a way to begin dealing with these frustrations and their consequences. Artifical insemination is a realistic alternative if a man is at risk, but of course is of no value to an at-risk female. Moreover, adoption is often thwarted by the adoption agency's concern for the eventual health of the at-risk adoptive parent. These kinds of considerations apply to the many similar autosomal dominant neurodegenerative disorders with adult onset and progressive deterioration. In the long run, genetic counseling for such disorders will be facilitated by presymptomatic direct diagnosis or indirect inference by linkage studies (see Chapter 9) applied to adults, children, or even fetuses through amniocentesis.

Autosomal Recessive Disorders

The *Meckel syndrome* has many diffuse manifestations that are tragically obvious at birth: occipital encephalocele, microcephaly, cleft lip and palate, polydactyly, congenital heart defects, and polycystic kidneys (21). If not stillborn, the infant usually dies in the neonatorum. Most instances of severe congenital malformations are due to chromosomal abnormalities, teratogens, or nonspecified developmental aberrations, but the Meckel syndrome exemplifies that multiple malformation disorders can have a single-gene etiology. When there are severe anomalies, therefore, a possible chromosome defect must be investigated by karyotyping, postmortem if necessary, and teratogens sought; if neither is found the possibility of a high recurrence risk has to be considered. The recurrence risk of the Meckel syndrome among siblings is 1 in 4, but this risk can be decreased sharply by using amniocentesis to determine amniotic fluid alpha-fetoprotein or it can be obviated by artificial insemination by donor.

Oculocutaneous albinism is obvious at birth and, as the name indicates, it involves the pigmented portions of the eye and skin (22). Its features include visual impairment and blindness, very pale skin, white or pale blond hair, and susceptibility to the adverse effects of ultraviolet light, such as sunburn and skin neoplasia. Originally considered a single disorder, it is manifested in at least two known varieties, each characterized by the presence or absence of tyrosinase activity in melanocytes. If a person with the tyrosinase-negative variety marries a person with the tyrosinase-positive type, none of their children will be affected with either type, indicating that the respective mutant genes reside at separate loci. This is an excellent example of genetic heterogeneity.

Infantile multicystic kidney disease is an isolated anomaly, and, for this reason, it is usually not recognized until renal failure occurs in later infancy or early childhood (23). It is important that the disease be distinguished from sporadic nongenetic renal dysplasia with cysts and from the occasional infantile presentation of adult-type polycystic kidney disease. Its 1 in 4 recurrence risk is an important consideration for subsequent pregnancies.

Cystic fibrosis is a relatively common autosomal recessive disorder that can be obvious at birth or manifested any time in childhood, or even escape detection until adulthood (24,25). It is primarily a disorder of exocrine glands of the skin, gastrointestinal tract including the pancreas, respiratory tract, and male reproductive tract; there is also anomalous development of the vas deferens. Presenting and long-term clinical problems include neonatal bowel obstruction (meconium ileus); malabsorption and pancreatic insufficiency; respiratory distress; inspissation of bronchial secretions and repeated respiratory infections; general failure to thrive; infertility in the male; and elevated concentrations of sodium, potassium, and chloride in eccrine sweat. These high concentrations are the basis for the one reliable laboratory method of detecting or documenting the disorder—the sweat test. Among whites, the incidence of cystic fibrosis is 1/1600 and the carrier (heterozygote) rate 1/20. The carrier rate among blacks is about 1/50, and for Orientals it is much smaller. Genetic counseling for this disorder is somewhat complicated because the carrier rate is so high, because affected males are generally sterile, and because there is no reliable method of carrier detection. A couple with an affected child has a 1 in 4 risk that each subsequent child will have the disorder. For every new case, all prior siblings should be evaluated, with at least one sweat test included, to identify mildly affected individuals. Unaffected siblings will have a 2 in 3 chance of being carriers; if they marry into the general white population (nonconsanguineously) where the carrier rate is 1 in 20, their chance of having an affected child is 1 in 120. Even though this ratio is low, it is still 13 times greater than that for the general population. For an affected woman the risk of having affected children would be 1 in 40. Artificial insemination by donor for parents of an affected child is useful, but less so than with rarer disorders, reducing the risk from 1 in 4 only to 1 in 80. When heterozygote (carrier) detection becomes available, it will thus have tremendous impact in reducing the recurrence risk in at-risk families. Prenatal diagnosis of this disorder should also accompany or closely follow the development of an accurate carrier detection test.

The *Usher syndrome* is an autosomal recessive disorder that is characterized by congenital sensory deafness and retinitis pigmentosa; some patients may also be mentally retarded and some develop cataracts (26). All cases of congenital deafness must be evaluated for this disorder, and the same is true of all cases of retinitis pigmentosa (see below). Making this specific diagnosis is important because it clarifies the inheritance pattern of its two major features, both of which may have other genetic or nongenetic etiologies.

Werdnig-Hoffman disease (infantile progressive spinal muscular atrophy) is almost uniformly lethal in early childhood, though it becomes apparent only after several months of apparent normalcy (27). The diagnosis should be established by muscle biopsy, since clinical differentiation from other disorders is not always possible if the manifestations are not typical, especially in terms of severity. Other progressive spinal muscular atrophies may have a somewhat different clinical course, though with overlap regarding time of onset.

Wilson disease (hepatolenticular degeneration) is manifested in various organ systems at various ages in children and adults (28). Early diagnosis is important; not only does the 1 in 4 recurrence among siblings become clear, but specific, highly effective therapy is available (penicillamine). The basic defect involving the transport and tissue accumulation of copper is reflected by decreased serum levels of copper and its transport protein ceruloplasmin, and increased urinary excretion of copper. Clinical features include Kayser-Fleischer rings in the perimeter of the corneas, liver failure and cirrhosis, and indications of basal ganglia dysfunction (i.e., Parkinsonism), all resulting from abnormal deposition of copper in the target organs. Of utmost importance is the amazing variation in severity and form of expression from one person to another, both within and between families. One child may present at an early age with neurological problems and a sibling years later with cirrhosis (or even under the guise of "chronic active hepatitis"). When a person is diagnosed, all siblings should therefore be evaluated for the disorder; this includes measurement of 24-hour urine copper levels, the most sensitive noninvasive means of assessment. Prenatal diagnosis is not yet possible, but artificial insemination by donor is a reasonable consideration.

Alpha-1-antitrypsin deficiency (AATD) may present at various ages, from infancy to adulthood, depending on the mutant alleles involved, environmental contributions (e.g., cigarette smoking), and other factors not yet understood (29). In its severest forms, clinical problems include neonatal hepatitis, childhood or adult cirrhosis, and distinctive chronic obstructive pulmonary disease and emphysema developing at an early age. There are more than 15 mutant alleles for this locus, each distinguished by the behavior of the protein (alpha-1-antitrypsin) on immunoelectrophoresis (Pi typing) and its ability to degrade substrate under standard conditions. Approximately 4% of the general population is said to be heterozygous for one of the mutant alleles. Currently, at-risk families are usually identified through an offspring with either the infantile or childhood-onset liver or lung involvement. At least three considerations make genetic counseling difficult: specific identification of the Pi type is not routinely available, the severity and form of the clinical manifestations have not been entirely consistent even within sibships, and there is controversy as to whether carriers (heterozygotes) for the more deleterious alleles may have associated hepatic or pulmonary morbidity (30). In any event, genetic counseling, based on knowledge of the specific Pi types, must inform parents of an affected child that there might be a recurrence risk as high as 1 in 4 for the same or a similar disorder. Prenatal diagnosis is not yet available, and when it is, care will have to be taken not to equate a given genotype with a definite expected clinical course or prognosis.

X-linked Recessive Disorders

Anhydrotic ectodermal dysplasia is a congenital disorder with diffuse manifestations (31). Characteristic features include a typical facies, short and fragile scalp hair, misshapen teeth, and absence of sweating (anhydrosis). Early diagnosis is necessary to avoid morbidity from periodic fevers or even heat prostration. Although findings can be subtle in infancy, this disorder's diagnosis may be facilitated by a positive family history of other affected males and partial expression in the boy's mother or other carrier females. The purpose of discussing the disorder here is precisely the latter point: its not infrequent expression in heterozygous females. This phenomenon is important for understanding X-linked recessive disorders in general. Thus, in the genetic counseling for this disorder a carrier female must not only be informed of the standard 50% risk for each son, but also of the possibility that there will be some manifestations in her carrier daughters. Likewise, an affected male would have to be informed that each of his daughters would have a substantial risk of manifesting her carrier status in some adverse manner.

X-linked aqueductal stenosis is an important disorder because of the relatively high frequency of males with hydrocephalus due to aqueductal stenosis (32). Clearly, all of them do not have the X-linked form of this condition, and it is impossible to be certain about the recurrence risks. The points to be considered are genetic heterogeneity, phenocopies, and new mutations. Congenital toxoplasmosis and other developmental insults can cause this deformity, introducing phenocopy mimicry into the picture. Genetic heterogeneity involves presumed polygenic as well as X-linked patterns. Finally, in the absence of a positive family history it is difficult, if not impossible, to put the X-linked risk in perspective because it is a genetic lethal (precludes procreation) and the mutation rate is unknown, as is the proportion of X-linked cases among all instances of aqueductal stenosis. The genetic considerations indicate that if an intrauterine insult has not been documented, the possibility of recurrence among future sons may be as high as 1 in 2. Because the polygenic variety almost certainly accounts for many of these cases, and because at least one-third of fresh X-linked mutations would be restricted to the affected boy, the cumulative risk for recurrence on an X-linked basis is quite low in the absence of a positive family history. In any event, because this inheritance pattern must be acknowledged to account for some males with the disorder, the counselor cannot give only the relatively lower, narrower recurrence risk, as when the proband is female. Whenever an X-linked lethal mutation must be considered in a heterogeneous grouping, this same caution in counseling is necessary.

Fabry disease (ceramide trihexosidase deficiency) has protean manifestations of childhood and adult onset (33,34). Its clinical features include unexplained episodes of fever, paresthesias, a distinctive skin lesion (diffuse angiokeratomas), heart failure, and kidney failure. When the skin lesions are present and are recognized, diagnosis is relatively straightforward. In some forms, however, the angiokeratomas are minimal or absent. Partial expression in carrier females is very common, though they may not develop skin lesions. Carrier detection and prenatal diagnosis are readily available. Affected males and carrier females have been treated by renal transplant, which provides both renal function and a source of the normal enzyme. So the general points to be appreciated are: more than one type of mutant allele (genetic heterogeneity), partial expression in carriers (heterozygote manifestation and Lyonization), carrier identification and prenatal diagnosis, and a model system for attempts at enzyme replacement therapy for a genetic disorder.

Duchenne muscular dystrophy (DMD) is an X-linked recessive disorder of childhood onset that precludes procreation and has an ultimately fatal outcome (35,36). It is characterized by the insidious onset of muscle weakness by age 3 to 5 years, progressive loss of strength and death in later childhood or early adulthood, pseudohypertrophy of the calf muscles in the early stages, massive elevations of serum muscle enzyme levels (especially creatine phosphokinase, CPK), and typical findings on muscle biopsy. Because it is a genetic lethal, new mutations play an important part in the origin of new cases. This disorder is a challenge to genetic counselors: its burden is tremendous for the patient and his relations so that prevention is especially important, carrier detection is less than satisfactory, specific prenatal diagnosis is not possible, and there are a number of other neuromuscular disorders which can be differentiated from DMD only with some difficulty. The counseling effort focuses on assessing the degree of risk for females who are potential carriers. There are basically two types of counseling situations. One, the less frequent, involves confirmation of a woman's carrier status and the consequent 50% risk of each subsequent son being affected and each daughter a carrier. The other situation involves determining the likelihood that an at-risk woman actually is a carrier; the magnitude of that risk; and the risk for her daughters, sisters, mother, and maternal aunts. This assessment is based on genetic principles (e.g., X-chromosome segregation, Lyonization) and empiric data, including the reliability of purported carrier detection methods. The point is that information derived from a variety of sources and from different vantage points can be utilized in a cumulative fashion to establish a risk figure that is usually much lower than that derived from genetic information alone. This Bayesian approach (37) will be applied to a specific family with DMD in Chapter 10 (p. 208). Suffice it to say here that the situation surrounding this disorder is always sufficiently complex and the burden sufficiently high that assistance through a genetic counseling clinic is warranted. The evaluation would include: 1. Documentation of the diagnosis as DMD. 2. Construction of a complete family pedigree to determine where the mutation has arisen and which family members have the greatest risks of being carriers, thereby establishing where to start individual investigations. 3. Individual investigation involving at least a neurological examination and three serial serum CPK determinations. It is incorrect to estimate recurrence risks solely on the basis of one affected proband and the degree of an individual's relatedness to him.

Hemophilia A (Factor VIII deficiency) is an X-linked recessive disorder of blood coagulation that becomes apparent at any time from birth on into childhood (38,39). There are relatively severe forms with bleeding from wounds or into tissues and joints and milder forms without much tissue or joint hemorrhage. In either case there is an absolute deficiency of coagulation Factor VIII, whether assessed in terms of coagulant activity or antigenic protein. The counseling problems are similar to those posed by families with Duchenne muscular dystrophy, but with two important differences: genetic fitness is much better (i.e., hemophiliac men not infrequently procreate) and carrier detection is much more reliable, relying as it does on both activity and antigen levels. As with DMD, prenatal diagnosis of the specific disease is not yet readily available. Thus, this mode of preventing either disorder entails determining fetal sex and considering abortion if it is male. Hemophilia B (Factor IX deficiency), also an X-linked trait, may be approached in the same way, although carrier detection is much less reliable. Taking into account the ability of men affected with hemophilia A to have children, it is critical to recognize that none of their sons will be affected or be carriers, and that all their daughters must be carriers (obligate heterozygotes). The daughters will thus have 50% risk of passing on the disorder to each of their sons. Moreover, carrier females may indeed have clinical problems with blood coagulation, which must be kept in mind when amniocentesis, surgery, or the administration of certain drugs (e.g., aspirin) is considered. Although the inheritance pattern of hemophilia has been recognized for over 2,000 years, many hemophiliac families are not given the benefit of this knowledge for use in family planning until after another affected son is born. Morbidity has decreased considerably with recent improvements in therapy, but hemophilia A is still a high-burden disorder responsible for substantial suffering and economic hardship.

The *testicular feminization syndrome* (TFS) results from the failure of embryologic end organs to respond to androgenic steroids despite the presence of a Y chromosome (40). Affected women thus have a 46,XY chromosome constitution, and their external anatomy is unquestionably feminine. As they also have primary amenorrhea and a blindly-ending vagina, however, they are sterile. In their management it is critical that the intraabdominal gonads be removed because of their propensity to lead to gonadoblastomas or dysgerminomas. Affected individuals in a family are related through unaffected females, although—as is typical of many X-linked recessive traits—carriers may have minor manifestations (e.g., sparse axillary hair, delayed menarche). These data, plus the known X-linkage of this trait in mice, establish the X-linked nature for humans even though affected humans never procreate. For this reason all sisters of affected persons should be evaluated to ascertain whether they are also affected and, if not, informed of their possible carrier status. Male siblings should be informed that they do not have the gene and cannot pass it on.

X-linked Dominant Disorders

Focal dermal hypoplasia is apparently lethal in males (41). In females, it is a congenital disorder with such characteristic features as cutaneous linear pigmentation, atrophy and localized defects of the skin through which fat often protrudes, and mucous membrane and cutaneous papillomas. Various digital, oral, and ocular abnormalities are also present in different forms and combinations. With X-linked dominant traits the ratio of affected females to males is 2:1, and males are expected to be affected more severely. The major distinction from X-linked recessive traits is that female offspring of females heterozygous for the mutant gene have a 1 in 2 risk of manifesting the disorder.

Albright hereditary osteodystrophy is considered to be an X-linked dominant trait, but similar syndromes with an apparently autosomal dominant pattern (i.e., male to male transmission and equal severity in males and females) tend to confuse the issue (6,7). The disorder is also known as pseudohypoparathyroidism or pseudopseudohypoparathyroidism(!), reflecting decreased and normal serum calcium levels, respectively. Parathyroid levels are normal, the endocrine aberration reflecting end-organ unresponsiveness. Clinical features include short stature, shortening of metacarpals and metatarsals, round facies, cataracts, and cranial and cutaneous calcifications. Brachydactyly, type E (42) is a relatively common autosomal dominant disorder manifested mainly by short metacarpals, metatarsals, and phalanges; and short stature; it is often misconstrued to be pseudopseudohypoparathyroidism. The distinction is important because of the different natural history of these two diseases. In addition, the X-linked trait cannot be transmitted from father to son, whereas brachydactyly, type E can.

Multiple Patterns

Retinitis pigmentosa is an extreme example of genetic heterogeneity (43,44,45). It is presented here because it is one of those conditions that occurs either as a part of a larger syndrome or as an isolated entity and because its inheritance pattern can be autosomal dominant, autosomal recessive, or X-linked recessive. When the diagnosis is made on the basis of abnormal retinal pigmentation and gradual circumferential restriction of vision, other associated features should therefore be considered, such as deafness (Usher syndrome, Alstrom syndrome), mental retardation, polydactyly, obesity, and hypogonadism (Laurence-Moon and Bardet-Biedl syndromes). When there is a clearly positive family history, counseling is straightforward. But genetic counseling is often sought when there is only one affected person. If the individual is a female and both parents are normal by history and fundus examination, autosomal recessive inheritance or a new autosomal dominant mutation are both likely and indistinguishable explanations. If the sole person affected with isolated retinitis pigmentosa is male, and again both parents are normal, the alternatives include autosomal recessive inheritance, a new autosomal dominant mutation in the patient, and an X-linked recessive mutation in the patient or his mother. The only thing that can be done under these circumstances is to acknowledge the dilemma and the fact that there is no way to resolve it.

I chose to end this chapter with such a dilemma to emphasize that there is more to genetic counseling than merely quoting risk figures once a diagnosis has been made. In addition to genetic data, many empiric, social, and other considerations enter into the process.

Cited References

1. Murphy EA, Chase GA: *Principles of Genetic Counseling*. Chicago, Year Book Medical Publishers, 1975.

2. Wachtel SS, Koo GC, Breg WR, et al: Expression of H-Y antigen in human males with two Y chromosomes. *New Engl J Med* 293:1070, 1975.

3. Lyon MF: X-chromosome inactivation and developmental patterns in mammals. *Biol Rev* 47:1, 1972.

4. Hathaway H, Lubs M, Kimberling W, et al: Carrier detection in classical hemophilia. *Pediatrics* 57:251, 1976.

5. Kousseff BG, Beratis NG, Danesino C, et al: Genetic Heterogeneity in fucosidosis. *Lancet* 2:1387, 1973.

6. Hermans PE, Gorman CA, Martin WJ, et al: Pseudohypoparathyroidism (Albright's hereditary osteodystrophy). A family study. *Mayo Clin Proc* 39:81, 1964.

7. Mann JB, Alterman S, Hill AG: Albright's hereditary osteodystrophy comprising pseudohypoparathyroidism and pseudo-pseudohypoparathyroidism, with a report of two cases representing the syndrome occurring in successive generations. *Ann Intern Med* 56:315, 1962.

8. Stern C: *Principles of Human Genetics* (3rd ed). San Francisco, WH Freemen, 1973, pp 296–302.

9. Jones KL, Smith DW, Harvey MAS, et al: Older paternal age and fresh gene mutation; Data on additional disorders. *J Pediatr* 86:84, 1975.

10. Spranger JW, Opitz JM, Bidder V: Heterogeneity of chondrodysplasia punctata. *Humangenetik* 11:190, 1971.

11. Gall JC Jr, Stern AM, Cohen MM, et al: Holt-Oram syndrome: Clinical and genetic study of a large family. *Am J Hum Genet* 24:24, 1972.

12. Goodman RM: A family with polysyndactyly and other anomalies. *J Hered* 56:37, 1965.

13. McKusick VA: *Heritable Disorders of Connective Tissue* (4th ed). St. Louis, C. V. Mosby, 1972.

14. Hall JG, Herrod H: The Stickler syndrome presenting as a dominantly inherited cleft palate and blindness. *J Med Genet* 12:397, 1975.

15. Brasfield RD, Das Gupta TK: Von Recklinghausen's disease. A clinicopathological study. *Ann Surg* 175:86, 1972.

16. Riccardi VM, Kleiner B: Neurofibromatosis; A neoplastic birth defect with 2 age peaks of severe problems. *Birth Defects Original Article Series* (in press).

17. Pope FM: Autosomal dominant pseudoxanthoma elasticum. *J Med Genet* 11:152, 1974.

18. Pope FM: Two types of autosomal recessive pseudoxanthoma elasticum. *Arch Dermatol* 110:209, 1974.

19. Hatfield PM, Pfister RC: Adult polycystic disease of the kidneys (Potter type 3). *JAMA* 222:1527, 1972.

20. Chandler JH, Reed TC, Dejong RN: Huntington's chorea in Michigan. *Neurology* 10:148, 1960.

21. Fitch N, Pinsky L: The Meckel syndrome with limited expression in relatives. *Clin Genet* 4:33, 1973.

22. Witkop CJ Jr, Nance WE, Rawls RF, et al: Autosomal recessive oculocutaneous albinism in man: Evidence for genetic heterogeneity. *Am J Hum Genet* 22:55, 1970.

23. Anand SK, Chan JC, Lieberman E: Infantile polycystic kidney disease and hepatic fibrosis in children. *Am J Dis Child* 29:810, 1975.

24. Bowman BH, Mangos JA: Current concepts in genetics; Cystic fibrosis. *New Engl J Med* 294:937, 1976.

25. Committee for Evaluation of Testing for Cystic Fibrosis, National Academy of Sciences: Evaluation of testing for Cystic Fibrosis. *J Pediatr* 88:711, 1976.

26. Kloepfer HW, Laquaite JK, McLaurin JW: The hereditary syndrome of congenital deafness and retinitis pigmentosa: (Usher's syndrome) *Laryngoscope* 76:850, 1966.

27. Byers RK, Banker BQ: Infantile muscular atrophy. *Arch Neurol* 5:140, 1961.

28. Strickland GT, Leu ML: Wilson's disease; Clinical and laboratory manifestations in 40 patients. *Medicine* 54:113, 1975.

29. Talamo RC: Basic and clinical aspects of the alpha-1-antitrypsin. *Pediatrics* 56:91, 1975.

30. Lieberman J: Alpha-1-antitrypsin deficiency and chronic pulmonary disease. *N Y State J of Med* 76:181, 1976.

31. Reed WB, Lopez DA, Landing B: Clincial spectrum of anhydrotic ectodermal dysplasia. *Arch Dermatol* 102:134, 1970.

32. Holmes LB, Nash A, Zu Rhein G: X-linked aqueductal stenosis: Clinical and neuropathological findings in two families. *Pediatrics* 51:697, 1973.

33. Wise D, Wallace HJ, Jellinek EH: Angiokeratoma corporis diffusum: A clinical study of eight affected families. *Quart J Med* 31:177, 1962.

34. Clarke JT, Knaack J, Crawhall JC, et al: Ceramide trihexosidosis (Fabry's disease) without skin lesions. *New Engl J Med* 284:233, 1971.

35. Skyring AB, McKusick VA: Clinical, genetic, and electrocardiographic studies of childhood muscular dystrophy. *Am J Med Sci* 242:534, 1961.

36. Roy S, Dubowitz V: Carrier detection in Duchenne muscular dystrophy. A comparative study of electron microscopy, light microscopy and serum enzymes. *J Neurol Sci* 11:65, 1970.

37. Murphy EA, Mutalik GS: The application of Bayesian methods in genetic counseling. *Hum Hered* 19:126, 1969.

38. Wintrobe MM, Lee GR, Boggs DR, et al (eds): *Clinical Hematology* (7th ed), Philadelphia, Lea & Febiger, 1974, pp 1161–1172.

39. Bennett B, Ratnoff O: Detection of the carrier state for classical hemophilia. *New Engl J Med* 287:342, 1974.

40. Stenchever MA, Ng ABP, Jones JK, et al: Testicular feminization syndrome. Chromosomal, histologic and genetic studies in a large kindred. *Obstet Gynec* 33:649, 1969.

41. Goltz RW, Henderson RR, Hitch JM, et al: Focal dermal hypoplasia syndrome: A review of the literature and report of two cases. *Arch Dermatol* 101:1, 1970.

42. Riccardi VM, Holmes LB: Brachydactyly, type E: Hereditary shortening of digits, metacarpals and long bones. *J Pediatr* 82:251, 1974.

43. Sunga RN, Sloan LL: Pigmentary degeneration of the retina: Early diagnosis and natural history. *Invest Opthal* 6:309, 1967.

44. Kobayashi F: Genetic study on retinitis pigmentosa. *Jap J Ophthal* 4:82, 1960.

45. Warburg M, Simonsen SE: Sex-linked recessive retinitis pigmentosa. A preliminary study of carriers. *Acta Ophthal* 46:494, 1968.

46. Warburg M: Genetic counseling in retinoblastoma. *Acta Ophthal,* Supp 123:110, 1974.

General References

A. Bergsma D (ed): *Birth Defects Atlas and Compendium.* Baltimore, Williams & Wilkins, 1973.

B. Goodman RM (ed): *Genetic Disorders of Man,* Boston, Little, Brown, 1970.

C. Smith DW: *Recognizable Patterns of Human Malformations* (2nd ed). Philadelphia, WB Saunders, 1976.

CHAPTER FIVE

GENETIC MECHANISMS OF DISEASE – POLYGENIC DISORDERS

Introduction

We have seen that a keynote of genetic disorders is predictability. Polygenic disorders share this feature although the predicted recurrence risks are usually lower than for single-gene disorders. To emphasize predictability the term "polygenic" is used here instead of "multifactorial," which suggests multiple factors, including environmental ones. The distinction is not one of exclusion, but of emphasis.

The purpose of this chapter is to present the general clinical features of polygenic traits (Table 5-1) and then consider a number of them in depth. As in the two preceding chapters, the need to rely on both empiric and genetic data will be emphasized.

Principles, Terminology, and Techniques

Many normal traits, such as adult height, and more than a dozen specific disorders are inherited in a polygenic manner (Table 5-2). The clinical picture represents the cumulative effect of many genes with small positive or negative influences. The exact number of contributing genes is not known, but in any case it is not a matter of simple addition and subtraction: there is also a threshold phenomenon. The net effect of the adverse genes will not be obvious unless it exceeds a critical threshold value beyond which the severity will be directly proportional to the number and degree of influence of the pathologic genes. The threshold capacity in each parent can be thought of as a water glass, equal in capacity (volume) to that of the other parent. The possible contribution of each parent to the trait in question is represented by the amount of water in each glass (Fig. 5-1). If the water level is below the brim, there is no spillage, i.e., the trait is not apparent in either

TABLE 5-1
A Brief Glossary for Polygenic Disorders

Multifactorial: When more than one factor is responsible for the development of a physiologic trait or clinical disorder; usually implies a combination of genetic and environmental factors.
Polygenic: When multiple genes with small, additive effects determine the presence of a trait, it is said to be polygenic; often used synonymously with "multifactorial" because of a presumed or known environmental component.
Proband: The affected individual whose disorder brings a family for genetic evaluation and counseling.
Relative, first-degree: Relative of the proband who shares 50% (half) of his/her genes; includes parents, siblings, and offspring.
Relative, second-degree: Relative of the proband who shares 25% (one-quarter) of his/her genes; includes grandparents, uncles, nieces, etc.
Relative, third-degree: Relative of the proband who shares 12.5% (one-eighth) of his/her genes; includes cousins, great-grandparents, etc.
Threshold: With regard to polygenic traits, that level of combined genetic and environmental influence at or beyond which the influence becomes overtly obvious.

parent. Each of their children would be represented by another glass of the same size. If each parent contributed more than half a glass of water (genes), the third glass would overflow (become clinically apparent). The overflow could not be blamed on any given portion of the water, but, rather, would result from the combined volume of the two glasses. In terms of predictability, it is obvious that if you know how much water each glass contains, you also know whether or not there could be an overflow.

Although this analogy shouldn't be pushed too far, it has a definite utility. It correctly suggests that we should not be surprised to find that the parents of a child with a polygenic disorder are themselves normal. The analogy also makes it clear that individuals already affected (full glass) would be more likely to have affected children than would the general population. Thus, increased recurrence risks for polygenic traits apply to all first-degree relatives of the proband, including siblings and offspring. Second-degree relatives (e.g., aunts, uncles, grandchildren) also have a significantly increased risk (Fig. 5-2).

Other characteristic features of polygenic traits in general include (Table 5-3) a preponderance in one sex or the other, and

TABLE 5-2
Common Polygenic Disorders

Allergies (atopic)	Hirschsprung disease
Cleft lip/palate	Hydrocephalus
Cleft palate	Hyperlipidemias
Club foot	Mental retardation, nonspecific
Congenital heart disease	Neural tube defects
Congenital hip dysplasia	Pyloric stenosis
Congenital scoliosis	Urinary tract malformations
Diabetes mellitus	

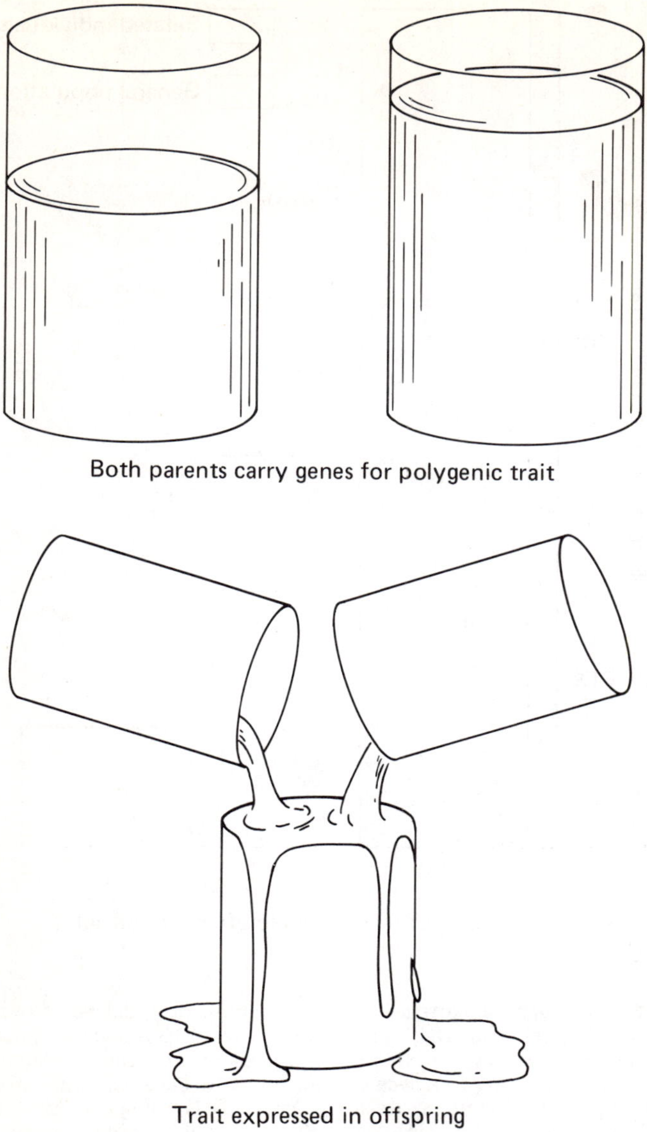

Figure 5–1 Water glass analogy for explaining polygenic inheritance.

increased risk for recurrence in proportion to the severity of the disorder in the proband and in proportion to the number of affected first- and/or second-degree relatives. Although environmental factors are thought to contribute to the development of polygenic defects, their nature and relation to genetic factors have not been clarified. But, when a polygenic defect develops in association with a specified environmental factor, it is at least prudent to avoid that factor in subsequent pregnancies.

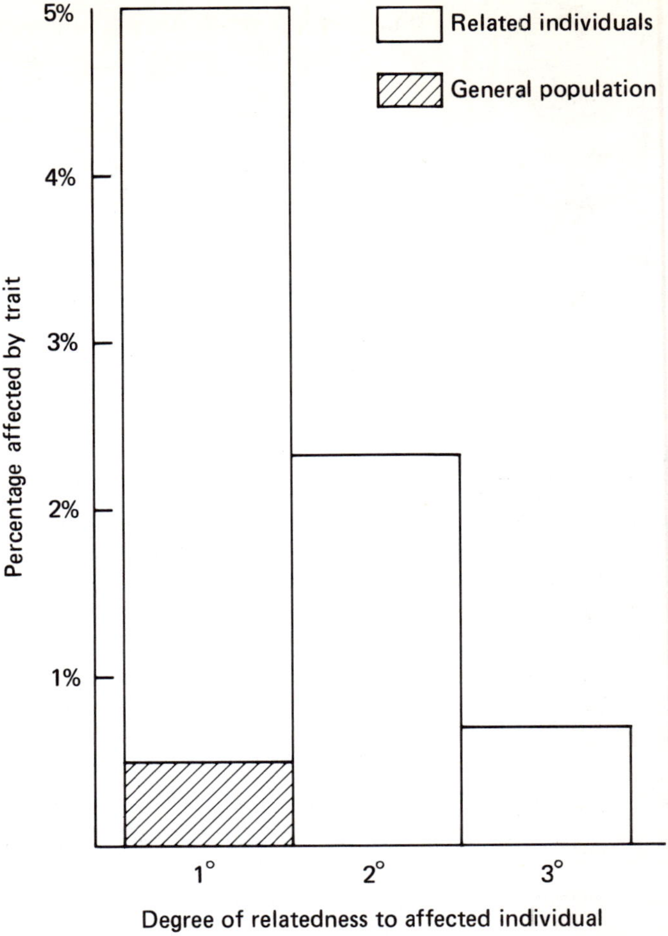

Figure 5–2 Polygenic recurrence risks as determined by degree of relatedness. For an idealized polygenic trait somewhere between 0.1% and 0.5% of the general population is affected, while the recurrence risks among blood relatives to the proband may be as high as 5% even with an otherwise negative family history. The 5% risk applies to first-degree relatives, and this is halved for second-degree relatives and quartered for third-degree relatives. Although the specific percentages might change slightly with specific disorders, the ratios and magnitudes of risk as a function of relatedness apparently remain relatively constant.

Polygenic congenital malformations are characterized morphologically by involvement of only a single organ system or tissue. There are two potential exceptions to this restriction, however: 1. If each of the defects derives from the same embryologic developmental field and thus they are nonspecifically associated as a part of the same "field defect;" for example, an upper-extremity anomaly may accompany congenital heart disease. 2. If the primary defect is sufficiently severe and critically located, there may be secondary defects, i.e., with a large, high, lumbar myelomeningocele there may be bladder, urethral, and anorectal anomalies as well as severely clubbed feet.

Three general rules can be derived from these anatomic considerations:

1. The risk of recurrence is for the same or a similar defect in the same organ system or tissue. A couple that has had a child with polygenic tetralogy of Fallot has an increased risk that another child will be born with congenital heart disease, but not cleft palate, club foot, etc.

2. When a child is born with an apparently isolated defect, a polygenic trait must be considered and other explanations ruled out. The other explanations might include teratogenesis (i.e., cleft palate due to diphenylhydantoin), a single-gene disorder (i.e., atrial septal defect as part of the Holt-Oram syndrome), or a nonspecific, nongenetic developmental anomaly (e.g., congenital amniotic bands). If the trait is indeed isolated and alternative explanations have been ruled out, then each of the defects discussed below should lead to genetic counseling based on the principles of polygenic inheritance. Isolated congenital defects other than those included in Table 5-2 should be considered to have a nonpolygenic etiology.

3. If there are two or more major defects, a nonpolygenic etiology should be considered, particularly if the defects derive from different developmental fields or one is not the consequence of the other. Under these circumstances chromosomal, single-gene, teratogenic, and unspecified developmental etiologies should be investigated early. Such an evaluation would include a careful family history and, especially if the central nervous system is involved, chromosome analysis and search for teratogens (e.g., serological screening, careful pregnancy drug history).

In the final analysis, genetic counseling for polygenic traits depends on the disorder being an essentially isolated defect for which previous empiric data have established the recurrence risks. The similarity in the incidence rates and empiric recurrence

TABLE 5-3
Principles of Recurrence Risks for Polygenic Disorders

1. Increased risk, compared to the general population, for recurrence of the disorder among first-, second-, and third-degree relatives of the affected person.
2. Baseline 2–5% risk for first-degree relatives to the proband; about half that risk for second-degree relatives.
3. Increasing risk with increasing numbers of affected genetic relatives, especially first- and second-degree.
4. Increased risk if affected person is of the sex less usually affected.
5. Increasing risk with increasing severity.

risks from one polygenic disorder to another reflects the similarity in the genetic mechanisms of their etiology. (But it does not in any way imply a common environmental contribution.) A generalization about these risks can be made as follows. Given an informative, negative family history except for the moderately affected newborn proband, whose sex is the one usually affected by the disorder in question, the recurrence risk is in the range of 2–5% for first-degree relatives. Factors that would favor the higher figure (5%) or frankly increase the recurrence risk include the following: presence of one or more similarly affected first- or second-degree relatives, occurrence in the less frequently involved sex, and extreme severity in the proband.

With these general principles in mind, we will now consider each of the well-known and well-established polygenic traits. Inclusion here does not preclude other etiologies in certain instances; on the contrary, alternative causes must be kept in mind. The following sketches are not intended to deal with all aspects of the disorders, just those relevant to counseling. It must also be remembered that incidence figures may vary substantially from one racial, ethnic, or national group to another.

Some Specific Disorders

Cleft palate as an isolated defect has an incidence of somewhat less than 1/1000 (1). It is seen more frequently in females than in males. It must be distinguished from the polygenic trait associated with cleft lip and from several autosomal dominant disorders, including the Stickler syndrome and the cleft lip/palate–lip pit syndrome. When cleft palate is associated with micrognathia and a floppy tongue, the combination is referred to as the Pierre-Robin malformation. This combination may itself be a sporadic occurrence or the manifestation of disorders such as the Stickler syndrome; as a secondary consequence of micrognathia, the palatal cleft assumes a rounded, horse-shoe configuration at the closed end, in contrast to the more typical, acutely angled "V" shape of a primary cleft palate (2). The cleft palate that can be associated with lip pits is not distinctive and the only evidence for its nonpolygenic nature is a positive family history or lip pits among first- and/or second-degree relatives (3). Other dominant disorders may also present as an isolated cleft palate. Therefore, the main point is that before any genetic counseling can be given, at least two things are critical: 1. delineation of a comprehensive family history and physical examination of the lips and palates, by inspection and palpation, of both parents. 2. A careful history of possible teratogens, particularly for drugs such as diphenylhydantoin, which have been implicated in the etiology of cleft palate.

This search for identifiable causes empahsizes a general principle in the approach to many polygenic disorders: The burden may be relatively small and/or the risk relatively low for the polygenic trait itself, but the burden and risk for the more complete single-gene syndrome which it partially phenocopies could be much greater.

Cleft lip, with or without cleft palate (CL/P) as a polygenic trait also has an incidence of about 1/1000 (1). It is one of the most common heritable traits for which genetic counseling is actively sought by an affected family. It is distinct from the previous disorder, which involves the palate alone, and is more common in males than females. In any given person CL/P may be manifest as an isolated bilateral cleft lip, an isolated unilateral cleft lip, or in association with varying degrees of clefting of the maxillary gingiva, hard or soft palates—cleft palate alone does not occur as part of this polygenic trait. In its most severe form, there is a bilateral cleft lip extending upward into the nostrils and backward through the gingiva and entire palate, hard and soft. Consequent problems include facial disfigurement, difficulty in sucking, difficulty in swallowing, nasal regurgitation, aspiration, increased susceptibility to middle-ear infections, and speech impairment. Proper therapy is best provided by a multidisciplinary team drawing specialists from plastic surgery, oral surgery, orthodontics, general dentistry, speech pathology, otolaryngology, audiology, psychiatry, social services, birth-defects clinics, and genetics.

This description of the full-blown syndrome is meant to emphasize how great the burden of the disorder can be. In particular, the severe psychosocial consequences of the facial disfigurement give CL/P a special place in genetic counseling. Parents often have unwarranted concerns about the child's ultimate mental abilities and misconceptions about the causes and the recurrence. Some believe that recurrence is extremely unlikely (i.e., "one chance in a million"); others believe that recurrence is almost inevitable.

There are milder forms of CL/P, of course, their severity varying considerably both within and between families. For a moderately affected male with an otherwise negative family history, the recurrence risk is in the range of 3–5%. Before ascribing even this risk in a specific case, however, other causes must be ruled out as far as possible. The cleft lip/palate–lip pit syndrome and teratogenesis (e.g., diphenylhydantoin) should certainly be considered and the absence of other anomalies documented.

Neural tube defects are another type of midline deformity that can have polygenic inheritance (4,5). (Unless otherwise qualified, the term "neural tube defect" (NTD) will refer to the polygenic trait.) The specific anomalies range from anencephaly through encephaloceles, myelomeningoceles (Fig. 5-3), and lipomeningoceles to certain forms of spina bifida occulta, each representing

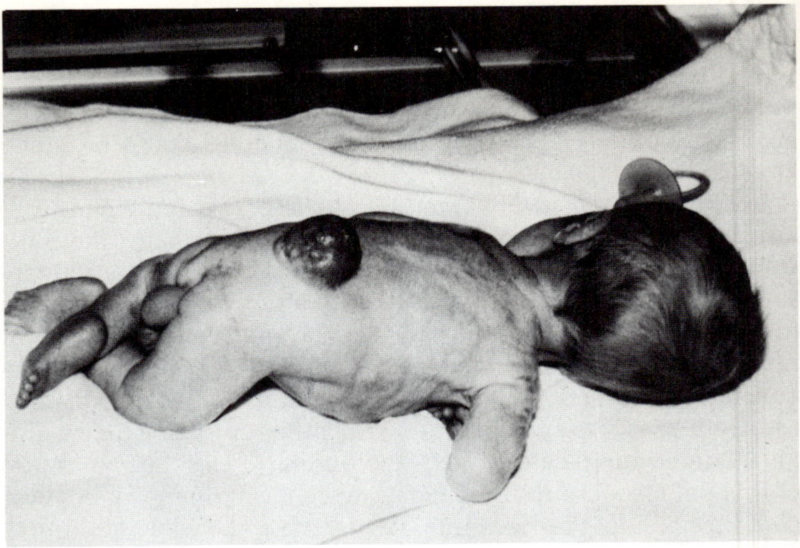

Figure 5–3 An infant with a large lumbar myelomeningocele.

different manifestations of the same basic genetic disorder. Each can also occur as a constant or occasional component of a broader syndrome: anencephaly or encephaloceles in the Meckel syndrome (autosomal recessive), lumbar myelomeningocele in the Patau syndrome (trisomy 13), occult spina bifida in the Warkany syndrome (see Chapter 3), and so on. In the absence of other anomalies, except perhaps cleft lip and palate, or those consequent to the neural defect, NTD should be considered polygenic in origin (5).

As with other polygenic traits, there is considerable variation in the incidence of NTD from one ethnic group to another, with a very high frequency in the Irish and a much lower one in Eastern Europeans, for example. The incidence in the United States is just under 1/1000 (6). But variable incidence rates do not automatically necessitate changing recurrence risks from one group to another. (Consider autosomal recessive traits, for example. Once parents have been identified as carriers, the risk is 1 in 4 regardless of the disease incidence in their native population.) For the present, baseline NTD recurrence risks for first-degree relatives may be estimated at 2–5% and for second-degree relatives, about 1% (see ref. 5). The possible role of specific teratogens is by no means established, and for the time being an isolated neural tube defect should be presumed polygenic even if there was gestational exposure to drugs or other agents. It is better to err on the side of polygenic etiology because of the potential prevention through prenatal diagnosis.

The burden of NTD, of course, depends on the type of defect, its location, and extent. It is very severe in two cases: anencephaly associated with stillbirth or neonatal death, which has an overwhelming but acute impact on the parents; and the child with a large, lumbar myelomeningocele who survives early vigorous medical and surgical therapy with subsequent chronic disability because he lacks sensation below the lesion, is paraplegic, develops hydrocephalus and consequently mental retardation, and requires urinary tract surgery and a colostomy to control excretion of body wastes. Such individuals may live into their late teens, are often institutionalized, and finally die of renal failure or sepsis. On the other hand, for about 1% of individuals with overt myelomeningoceles, aggressive treatment may afford a productive and satisfying life. This small group makes it impossible to generalize about the burden of lumbar myelomeningocele and difficult to make decisions about how aggressive early therapy (surgery especially) should be. Guidelines for making these decisions have been formulated (7). They are based on extensive experience and humane sensibility, but are not easy to apply, especially if everyone involved in a given case does not subscribe to them. In any event, genetic counseling for NTD is often difficult because of the wide range of primary defects and the variable consequences of lumbar myelomeningoceles.

Genetic counseling for NTD is further characterized by the role of prenatal diagnosis utilizing amniocentesis for alpha-fetoprotein determinations, ultrasound, and/or special radiographic techniques (amniography) (8,9). The details will be discussed in Chapter 8. In the long run, a positive family history of NTD may become the most common indication for prenatal diagnosis by amniocentesis. In addition, many genetic centers do amniotic-fluid alpha-fetoprotein determination on all prenatal diagnosis samples as a screening device (10). The role of screening pregnant mothers' serum for increased alpha-fetoprotein levels is still to be defined (11).

Because prenatal diagnosis and subsequent intervention are available, it is even more important to clarify the risks of NTD recurrence. The presence of occult spina bifida may enter into the estimate of risk (12). If the family history is already positive for NTD, occult spina bifida other than in segments L4, L5, and S1 may add to the recurrence risks. With a negative family history the picture is not as clearcut, but in at least some cases, occult spina bifida (not in L4, L5, or S1) associated with other vertebral defects may warrant assignment of NTD recurrence risks.

The syndrome of sacral agenesis (caudal dysplasia, caudal regression syndrome) is not ordinarily placed in the polygenic NTD spectrum (13). I think that it should be, and that parents of such children should therefore be counseled as for NTD in general and prenatal diagnosis should be used. My reasons for this opinion are twofold: 1. It could be argued by analogy that sacral agenesis is the caudal equivalent of anenencephaly. 2. Children are known to have sacral agenesis, lumbar myelomeningocele, and encephaloceles together, implying a common pathogenesis if not a common etiology.

In summary, polygenic NTD has a wide range of manifestations (including occult spina bifida), a frequently high burden with substantial morbidity and mortality, and definitely increased recurrence risk among first- and second-degree relatives. And prenatal diagnosis is available.

Hydrocephalus in isolation may be inherited as a polygenic trait with an incidence of about 1/1000 (14). In clarifying recurrence risk, the problem is to rule out single-gene traits, infectious teratogens, and postnatal causes such as subarachnoid hemorrhage, trauma, and meningitis. It is also important to rule out other types of head enlargement such as cerebral gigantism and heritable megalencephaly. Our ability to readily distinguish hydrocephaly from hydranencephaly and other types of cerebral enlargement has been greatly enhanced by new low-dosage, computerized radiographic techniques (computerized axial tomography) (15).

If hydrocephalus is present at birth, the main thing to rule out is an infectious teratogen, especially toxoplasmosis. This can usually be done by serologic evaluation of the infant and mother (serum antitoxoplasmosis antibodies) plus examination of the child for retinal lesions and choroid plexus calcification. In either case, it is then important to determine whether there is stenosis of the aqueduct of Sylvius since that lesion in males can represent an X-linked recessive trait. If the onset of hydrocephalus is postnatal, the differential diagnosis broadens but still includes the polygenic trait. If other causes of the hydrocephalus have been excluded as far as possible, polygenic recurrence risks, with a baseline magnitude of 2–5% for first-degree relatives, would apply.

The most straightforward situation involves an infant with isolated congenital hydrocephalus of the communicating type and a negative teratogen workup. With an otherwise negative family history, a polygenic etiology would be presumed. If it is a female infant with an anomalous aqueduct of Sylvius and the other details unchanged, the counseling would be essentially the same. For a male infant with congenital hydrocephalus associated with aqueductal stenosis (and the other details unchanged), one cannot ignore the possibility of the X-linked form of this disorder. The genetic counseling would involve about the same numerical risk, but with the caveat that the actual risk may be higher, though it is indeterminate at present. The uncertainty arises from our ignorance as to what proportion of aqueductal stenosis is due to the X-linked trait and from the fact that any new X-linked mutation could have occurred in the affected child himself, thus carrying no risk for siblings. This uncertainty magnifies the need to obtain an accurate and complete family history and to avoid an overly optimistic prediction for future siblings. On the other hand, it is unreasonable to perform amniocentesis for fetal sex determination and then consider abortion, unless an X-linked pattern is already established. The only useful method of prenatal diagnosis at present is ultrasound assessment of head size, but the results are usually difficult to interpret and the hydrocephalus may not become apparent until late in pregnancy, or even after birth. Amniotic fluid alpha-fetoprotein determinations have not been useful in detecting hydrocephalus in the absence of neural tube defects.

Congenital heart disease (CHD) is quite common as a polygenic trait; its incidence is about 6–8/1000 (16). The term encompasses relatively simple defects, such as ventricular septal defect (VSD) and atrial septal defect, and complex defects such as tetralogy of Fallot and transposition of the great vessels. Each form of CHD can also occur as a component of hundreds of other disorders of all etiologic classifications (Chapter 2): chromosomal disorders including the Down syndrome (Ebstein anomaly) and X-chromosome monosomy (aorta coarctation), single-gene disorders (Holt-Oram syndrome, Noonan syndrome, Ellis-Van Creveld syndrome, etc.), teratogenic abnormalities (Rubella syndrome, hydantoin syndrome), and developmental disorders for which the etiology and pathogenesis remain obscure (e.g., the Williams elfin-facies syndrome and many unnamed conditions). CHD is a cardinal feature of a number of single-gene disorders, such as Holt-Oram syndrome and other specific hand–heart syndromes. And in several single-gene disorders it is the sole abnormality, though they usually do not involve gross anatomic deficits—idiopathic hypertrophic subaortic stenosis (17), congenital heart block (18), familial Wolfe-Parkinson-White syndrome (19), etc.

As with the other polygenic traits, prompt evaluation to rule out more comprehensive diagnoses and identify nonpolygenic etiologies is critical for accurate genetic counseling. Four special considerations must be taken into account, however:

1. There may be associated, nonspecific, upper-extremity anomalies ("field defects"). Upper-extremity defects are likely to be found in association with CHD by virtue of the close topo-

graphic and temporal relationships of the developing limb and heart. It is as though the aberrant process "spills over" into adjacent and concurrently developing tissues.

2. There are additional problems which may represent secondary consequences of the heart defect itself. These include short stature, early developmental delay, and digital clubbing.

3. It may be difficult to decide whether transient defects belong in the same risk grouping, especially if the defect represents persistence of a prenatal physiologic state, such as a patent ductus arteriosus (PDA) in an infant with respiratory distress or born at a high altitude. These problems of neonatal persistence probably do not reflect the ordinary CHD picture, and parents should know that the usual CHD risks may not apply. Unfortunately, many infants with a physiologically persistent PDA are given nosologic classifications (e.g., computer-based discharge diagnoses) that do not make or imply the distinction.

4. With such a large number of affected persons, there must be substantial etiologic and pathogenetic heterogeneity. Further, the recurrence risks for very rare defects largely reflect projection, by analogy, of the empirically determined risks for the more common aberrations.

CHD affects males and females with about equal frequency. There are basically two types of recurrence problems: recurrence of the same defect and recurrence of CHD in general. If a young couple has had a child with a VSD, the risk for recurrence of a VSD is put at about 3%, while if the child had a complex and very unusual defect such as transposition of the great vessels, the quoted recurrence risk of 2–3% is for CHD in general. Even a 3% risk is only four or five times the background risk for the general population (in contrast, for example, to the 50-fold increase for the recurrence of neural tube defects, where the risks may go from 1/1000 to 1/20). From a genetic point of view, then, the main emphasis is on recurrence of CHD in general, and recurrence of the same lesion is a secondary consideration. This approach is supported by experience with many families and by data from veterinary medicine (20).

Because the recurrence risks for polygenic CHD when only one person in the family is affected are so little different from the general population risks and because prenatal diagnosis is not applicable, genetic counseling usually emphasizes secondary prevention. That is, in view of the four- to fivefold increased risk, each subsequent sibling and offspring to the proband should be carefully examined for CHD at birth, in the weeks that follow, and up to at least one year of age. An analogy is often drawn to safety belts in automobile driving: counseling and seat belts signal an acknowledgment of increased risk. It is also important to examine previously born siblings and parents for occult CHD, since positive findings might substantially increase recurrence risks.

Club foot is seen more often in males than in females; its overall incidence is about 1/1500 (21). The genetic counseling for unambiguous cases is straightforward and consistent with the general approach outlined above. There are, however, several special considerations regarding this disorder which are worth mentioning:

1. The most important of these is the need to distinguish the isolated, primary polygenic disorder from secondary defects associated with many chromosome abnormalities, certain single-gene disorders, and a number of nonspecific developmental conditions including congenital amniotic bands (22) and the Potter syndrome (23).
2. The last example noted raises an unresolved dilemma—whether club foot represents an intrinsic or a superimposed positioning defect (24). It should be emphasized, however, that whether the primary defect is intrinsic or extrinsic to the lower extremity, for the polygenic trait the same recurrence risks apply.
3. What is to be included under the designation "club foot" is not always clearly defined. For example, is metatarsus adductus a relatively minor, completely distinct anomaly, or is it merely a very mild form of club foot? Should the term "club foot" be used only for a talipes equinovarus deformity?

Congenital hip dysplasia, or congenital dislocation of the hip, has an overall incidence of about 1/1000 and a more frequent occurrence in females (25). As with club foot, there is controversy as to whether the primary defect is intrinsic or extrinsic to the hip. In either case, the same recurrence risks, based on general polygenic principles, apply. Moreover, since the secondary prevention efforts can be so effective, early recognition aided by an appreciation of the recurrence risks can substantially decrease the morbidity of this disorder.

Congenital scoliosis is the final skeletal polygenic disorder to be considered (26). This spinal curvature defect differs from most other congenital defects in that its presence is often not apparent until some years after birth. For this reason, frequent evaluation of siblings and offspring to the proband is important for early diagnosis and optimal preventive therapy. Congenital scoliosis is much more frequent in females, for whom it can become an important obstetrical problem, even necessitating cesarean section for delivery. The onset of clinical problems is usually insidious, which, in effect, confirms the point that early recognition can minimize the consequences. A moderately affected female with an otherwise negative family history has a 3–5% risk for having affected siblings and offspring. A male suffering an early-onset, severe disorder would have at least a 5% risk for recurrence among first-degree relatives. It is very important to rule out other causes of scoliosis, a task perhaps made somewhat more difficult by the possible delay in manifestations and the need to rule out such primary disorders as neurofibromatosis and the Marfan syndrome.

Pyloric stenosis is a polygenic congenital disorder with an overall incidence in whites of about 1/500, and a higher incidence in male infants (27). It may be recognized at the birth examination by palpation of an epigastric mass, or it may not be appreciated until the child continually regurgitates feedings. Though it may be part of a larger disorder, such as the Down syndrome, the defect is usually an isolated anomaly. Siblings and offspring of an affected male have a 3–5% chance of being affected, while the first-degree relatives of an affected female would have about a 15% risk. A positive family history would, of course, also increase the recurrence likelihood. Again, a major function of genetic counseling is to enhance the likelihood of early diagnosis through appreciation of the recurrence risk.

Hirschsprung disease, or congenital aganglionic megacolon, involves the absence of the myenteric plexus in segments of the terminal large bowel (28). There are two forms of the disorder: the short-segment form involves the rectum and sigmoid colon (80% of cases), while in the long-segment form the entire colon and even terminal ileum are aganglionic (20%). Males are affected more frequently, the male to female ratio depending on the length of bowel involved: rectum only—7.2:1; rectosigmoid—4.2:1, and long-segment form—1.5:1 (29). The long-segment form is probably not a separate disorder but the more severe expression of a single polygenic trait, even though the length of bowel involved in various members of a given family tends to be more or less constant. Recurrence risks are higher for the long-segment (1 in 8) than short-segment form (1 in 32). In addition, with the short-segment form the recurrence rate is higher for males (1 in 19 versus 1 in 145), while for the long-segment form the sex ratio is about equal. In any case, the baseline recurrence risk is about 3% for first-degree relatives, increasing to about 12% with the long-segment form.

Early diagnosis should follow an appreciation of these recurrence risks. Therapy commonly involves surgical removal of the aganglionic segment to overcome the severe constipation or obstipation that results from the inability of the defective bowel to evacuate its contents. As a polygenic trait this condition is an isolated defect, without consistently associated features to facilitate diagnosis. But, aganglionic colon occasionally occurs as part of the Down syndrome or another disorder.

Mental retardation of mild to moderate severity and nonspecific nature can be a polygenic trait (30). The precise incidence isn't known, but it is considered a relatively common type of mild or moderate retardation (i.e., IQ of about 70). Affected individuals are distinguished only by their retardation; they are anatomically and chromosomally normal, and show no specific or consistent biochemical aberrations. Before genetic counseling can proceed, it is therefore necessary to obtain an accurate family history and general history, conduct a detailed physical examination, perform a chromosome analysis, and screen for biochemical disorders such as phenylketonuria. The risks for offspring to such individuals must take into account the intellectual ability of the spouse since the offspring's ability will be, on the average, the mean of the parents' abilities. It is not uncommon for a couple, each with nonspecific polygenic mental retardation, to inquire about the prospects for the children they may have. Such couples may already be aware that their children will not have normal intelligence, but may fear that, in contrast to them, the impairment could be severe. Reassurance that this is not to be expected can be pivotal for their decisionmaking. It is, of course, important that additional counseling be provided with regard to the possible need for assistance with prenatal care, early child care, and enhancement of the home environment.

The key point here, however, is that the recurrence risk has a different character: Instead of representing the likelihood of an offspring superseding a given threshold and thereby manifesting the trait (i.e., 5% risk for cleft lip/palate versus 95% for no defect), the risk for the polygenic trait of intellectual impairment is virtually 100%, the degree of expression varying as a direct function of the parents' intellectual impairment. That is, the expected level of the child's intellectual performance on the average will be the mean of the parents' levels. It is beyond the scope of this discussion to speculate on whether the child's ultimate intellectual capacity is determined by strictly genetic factors or is also influenced by the domestic environment in which the child is raised. The empiric risk stands, nonetheless.

Urinary tract malformations sometimes occur in several members of a family (31). Because they are the most common type of human malformation, however, familial occurrence in at least some cases may be coincidental. In addition, some of these defects are of no consequence and remain undetected until found incidentally at surgery or autopsy (e.g., bifid ureter) and in some cases an autosomal dominant inheritance pattern may be involved (32), falsely skewing the expectations toward the higher side for the population at large. In practical terms, when a urinary tract anomaly has been found in an individual, other family members, current and future, are sufficiently at risk to necessitate obtaining a careful family history; screening intravenous pyelograms should be considered for first-degree relatives. This approach is warranted by the considerations that clinically significant malformations may remain occult for long periods, that surgical intervention could minimize, if not prevent, consequent morbidity, and that the presence of two or more affected first-degree relatives would substantially increase the recurrence risks. Because of the concurrent embryologic development of the urinary and auditory systems, assessment of hearing may also be in order.

Diabetes mellitus almost certainly represents more than one pattern of inheritance (33), including autosomal dominant, autosomal recessive, and polygenic forms. Taking into account the high incidence of diabetes mellitus and its variability, even within families, it is difficult to know in individual cases and families which of the following mechanisms, or combinations thereof, apply: new mutation, dominance with variable expression and variable penetrance, recessiveness and a high carrier rate and some degree of heterozygote manifestation, polygenic inheritance, and phenocopy. In addition, a distinction has to be made between the "juvenile onset" and "adult onset" forms of the disorder. Thus, when diabetes mellitus is present in a family the chances of its appearance in other family members depend on their relation to the proband, the time of onset and severity in the proband, and the number of other affected relatives. Because of all these vagaries, it is almost impossible to generalize. But with an accurate family history, individuals in counseling can be given reasonable risk estimates based on empiric data and genetic principles (34).

Essential hypertension (35), *arteriosclerotic heart disease* (36), *atopic allergies* (37), *schizophrenia* (38), and most instances of familial *hypercholesterolemia* (39) are considered to be polygenic traits. This classification has little relevance for genetic counseling in view of the late onset of these disorders and the role of multiple environmental factors. Respect for their polygenic nature, however, should prompt at-risk family members to have routine clinical and laboratory evaluations.

Connective tissue disorders such as systemic lupus erythematosus, rheumatoid arthritis, ankylosing spondylitis, and scleroderma are sometimes familial (40). Clear evidence for a genetic basis, however is usually lacking. Nevertheless, the demonstration of an association between specific genetic immunologic markers (the HLA system) and certain disorders (e.g., HLA-W27 and akylosing spondylitis) has lent credence to the hypothesis of at least a partial genetic component in these disorders (41).

Cited References

1. Fraser FC: The genetics of cleft lip and palate. *Am J Hum Genet* 22:336, 1970.

2. Hanson JW, Smith DW: U-shaped palatal defect in the Robin anomalad: Developmental and clinical relevance. *J Pediatr* 87:30, 1975.

3. Van der Woude A: Fistula labii inferioris congenita and its association with cleft lip and palate. *Am J Hum Genet* 6:244, 1954.

4. Carter CO, Laurence KM, David PA: The genetics of major central nervous system malformations, based on the South Wales socio-genetic investigation. *Develop Med Child Neurol* (Suppl) 13:30, 1967.

5. Holmes LB, Driscoll SG, Atkins L: Etiologic heterogeneity of neural tube defects. *New Engl J Med* 294:365, 1976.

6. Center for Disease Control: Time trends in anencephaly and spina bifida. *Congenital Malformation Surveillance July 1974–June 1975,* issued December 1975, Atlanta, pp 1–6.

7. Lorber J: Selective treatment of myelomeningocele: To treat or not to treat? *Pediatrics* 53:307, 1974.

8. Brock DJH: Prenatal diagnosis—chemical methods. *Br Med Bull* 32:4, 1976.

9. MacVicar J: Antenatal detection of fetal abnormalities—physical methods. *Br Med Bull* 32:4, 1976.

10. Milunsky A, Alpert E: Routine testing for alpha-fetoprotein in amniotic fluid. *Lancet* 1:1015, 1976.

11. Cowchock FS, Jackson LG: Diagnostic use of maternal serum alpha-fetoprotein levels. *Obstet Gynec* 47:63, 1976.

12. Wynne-Davies R: Congenital vertebral anomalies: Etiology and relation to spina bifida cystica. *J Med Genet* 12:280, 1975.

13. Blumel J, Evans B, Eggers GWN: Partial and complete agenesis or malformations of the sacrum with associated anomalies. *J Bone Joint Surg* 41A:497, 1959.

14. Warkany J: Hydrocephaly. in *Congenital Malformations—Notes and Comments.* Chicago, Year Book Medical Publishers, 1971, pp 217–231.

15. Harewood-Nash DC, Breckbill DL: Computed tomography in children: A new diagnostic technique, *J Pediatr* 89:343, 1976.

16. Perloff JK: *The Clinical Recognition of Congenital Heart Disease.* Philadelphia, WB Saunders, 1970.

17. Clark CE, Henry WL, Epstein SE: Familial prevalence and genetic transmission of idiopathic hypertrophic sub-aortic stenosis. *New Engl J Med* 289:709, 1973.

18. Schaal SF, Seidensticker J, Goodman R, et al: Familial right bundle-branch block, left axis deviation, multiple heart block and early death. *Ann Intern Med* 79:63, 1973.

19. Schneider RG: Familial occurrence of Wolff-Parkinson-White syndrome. *Am Heart J* 78:34, 1969.

20. Patterson DF, Pyle RL, Buchanan JW: Hereditary cardiovascular malformations of the dog. *Birth Defects Original Article Series,* vol VIII, No 5, pp 160–174, 1972.

21. Stewart SF: Clubfoot: Its incidence, cause and treatment: anatomical-physiological study. *J Bone Joint Surg* 33A:577, 1951.

22. Jones KL, Smith DW, Hall BD, et al: A pattern of craniofacial and limb defects secondary to aberrant tissue bands. *J Pediatr* 84:90, 1974.

23. Passarge E, Sutherland JM: Potter's syndrome. Chromosome analysis of three cases with Potter's syndrome or related syndromes. *Am J Dis Child* 109:80, 1965.

24. Dunn PMN: Congenital posturing deformities. *Br Med Bull* 32:71, 1976.

25. Carter CO, Wilkinson JA: Genetic and environmental factors in the etiology of congenital dislocation of the hip. *Clin Orthop* 33:119, 1964.

26. Wynne-Davies R: Familial idiopathic scoliosis: A family survey. *J Bone Joint Surg* 50A:24, 1968.

27. Carter CO: Genetics of infantile pyloric stenosis. *The Clinical Delineation of Birth Defects. XIII. GI Tract Including Liver and Pancrease*. Baltimore, Williams and Wilkins, 1972, pp 15–21.

28. Passarge E: The genetics of Hirschsprung's disease. Evidence for heterogeneous etiology and a study of sixty-three families. *New Engl J Med* 276:138, 1967.

29. Bodivan M, Carter CO: A family study of Hirschsprung's disease. *Ann Hum Genet* 26:261, 1963.

30. Reisman LE, Malheny AP Jr: *Genetics and Counseling in Medical Practice*. St. Louis, CV Mosby, 1969, pp 189–192.

31. Bois E, Feingold J, Benmaiz H, et al: Congenital urinary tract malformations: Epidemiologic and genetic aspects. *Clin Genet* 8:37, 1975.

32. Dwoskin JY: Sibling uropathology. *J Urol* 115:726, 1976.

33. Zonana J, Rimoin DL: Current concepts in genetics: Inheritance of diabetes mellitus. *New Engl J Med* 295:603, 1976.

34. Fuhrmann W, Vogel F: *Genetic Counseling* (2nd ed), New York, Springer-Verlag, 1976, pp 96–98.

35. McKusick VA: Genetics and the nature of essential hypertension. *Circulation* 22:857, 1960.

36. Slack J, Evans KA: The increased risk of death from ischemic heart disease in first degree relatives of 121 men and 96 women with ischemic heart disease. *J Med Genet* 3:239, 1966.

37. Lubs M-LE: Empiric risks for genetic counseling in families with allergy. *J Pediatr* 80:26, 1972.

38. Reich W: The schizophrenic spectrum: A genetic concept. *J Nerv Mental Dis* 162:3, 1976.

39. Lees RS, Wilson DE, Schonfeld G: The familial dyslipoproteinemias. *Prog Med Genet* 10:1973.

40. Arnett FC, Schulman LE: Studies in familial systemic lupus erythematosus. *Medicine* 55:313, 1976.

41. Arnett FC Jr.: The implications of HL-A W27. *Ann Intern Med* 84:94, 1976.

General References

A. Bonaiti-Pellie C, Smith C: Risk tables for genetic counseling in some common congenital malformations. *J Med Genet* 11:374, 1974.

B. Carter CO: Multifactorial genetic disease. in McKusick VA, Claiborne R (eds), *Medical Genetics*. New York, HP Publishing, 1973, pp 199–210.

C. Carter CO: Genetics of common single malformations. *Br Med Bull* 32:21, 1976.

D. Holmes LB: Inborn errors of morphogenesis. *New Engl J Med* 291:763, 1974.

E. Holmes LB: Current concepts in genetics: Congenital malformations. *New Engl J Med* 295:204, 1976.

F. Hook EB, Marden PM, Reiss KP, et al: Some aspects of the epidemiology of human minor birth defects and morphological variants in a completely ascertained newborn population (Madison study). *Teratology* 13:47, 1976.

G. Smith C, Mendell NR: Recurrence risks from family history and metric traits. *Ann Hum Genet* 37:275, 1974.

H. Warkany J: *Congenital Malformations—Notes and Comments.* Chicago, Year Book Medical Publishers, 1971.

CHAPTER SIX

BIOCHEMICAL DISORDERS

Introduction

Genetically determined biochemical disorders are also referred to as inborn errors of metabolism. Each produces characteristic alterations in tissue or cellular metabolism. The *primary defect* for each disorder is a heritable alteration in a molecule that either mediates a specific biochemical process or serves as a structural element within or between cells. The primary defect ultimately provides the basis for specific diagnosis and therapy. It is this specificity that distinguishes many biochemical disorders from most other genetic disorders (except those due to immunologic incompatibilities such as ABO or Rh transfusion reactions).

If a patient is thought to have a biochemical disorder, the task is then to prove it by demonstrating either the primary defect through a specific assay or abnormal levels of specific metabolites in blood, excretions, or solid tissue. Two practical considerations complicate this task: 1. The equipment and expertise required to make many specific diagnoses are not widely available. 2. For each disorder with well-defined laboratory criteria, there are usually several variant forms with subtle biochemical differences. Add to this the rarity of most metabolic disorders, and it is easy to see why a correct diagnosis can be frustratingly elusive or may be missed.

The purpose of this chapter is fourfold:

1. (Primarily) to give the reader an idea of what is meant by the term "inborn error of metabolism" or its equivalents. Clarifying the distinctions and overlap between "biochemical" and "nonbiochemical" disorders should make the diagnosis and implications of each type clearer.
2. To elucidate the general genetic principles relevant in common clinical situations. These include such concepts as mutation, heterozygote, compound heterozygosity, and genetic heterogeneity.

3. To indicate what is involved in the evaluation of a patient suspected of having a biochemical disorder. Such procedures introduce screening tests, employed routinely and uniformly in large populations or based on a specific rationale in individual cases.

4. To reiterate the principle that a genetic diagnosis can have immediate and long-range implications for the management of the affected person and the prevention of recurrences.

Individually, these disorders are usually rare, with incidences of much less than 1/1000. As a group they are not rare, but have a cumulative prevalence of 1–2% and a substantial impact on hospital and other health care resources.

Characteristics of Biochemical Disorders

All single-gene (Mendelian) disorders ultimately will be accounted for in molecular terms. For many of them at this time, however, virtually nothing is known about the nature of the biochemical defect involved in their pathogenesis. The Meckel syndrome, neurofibromatosis, and polysyndactyly fall into this category. For others, the general type of biochemical defect is suspected, but efforts to specify it have failed. Achondroplasia, cystic fibrosis, and Duchenne muscular dystrophy are examples of this group. For a relatively small number of the 2000 single-gene disorders recognized, the biochemical defect underlying each condition is so well known that it also serves as a *sine qua non* criterion for making the diagnosis. These disorders have an importance quite disproportionate to the relative rarity of most of them. First, by characterizing the defect down to a minute level, we have learned a great deal about the nature of genes and mutations. For example, the discovery that the hemoglobin of sickle-cell anemia (1) differs from normal adult hemoglobin by only one amino acid helps us understand that a mutation need involve only the smallest corresponding component of the gene itself. Second, the mechanisms involved in one disorder can often be projected to other more common disorders. For example, the demonstration of a biochemical basis for the overproduction of uric acid in the rare Lesch-Nyhan syndrome (2,3) has contributed much to our understanding of and ability to treat gouty arthritis. Third, for a family in which the disorder occurs, recognition of the primary defect may afford treatment and/or the potential of prevention through prenatal diagnosis in subsequent pregnancies. For example, the consequences of phenylketonuria (4) and maple syrup urine disease (5) can be minimized by dietary restrictions; and the recurrence of the latter can be prevented through the use of amniocentesis for prenatal diagnosis.

Before considering biochemistry *per se,* let us review some general clinical features of inborn metabolic disorders (see also Table 6-1). Genetically determined metabolic disorders should be suspected in four sets of circumstances:

1. When the patient's signs and symptoms specifically suggest a certain type of metabolic disorder or even a particular diagnosis. For example, a partially masculinized female newborn that develops dehydration and shock 10–20 days after birth points to the adrenogenital syndrome (6); a boy with mental retardation, athetosis, self-mutilation and hyperuricemia points to the Lesch-Nyhan syndrome; and a male infant that fails to thrive and has kinky hair points to Menkes disease (7).
2. When an otherwise healthy child or adult develops diffuse or tissue-specific lesions, especially central nervous system deterioration, that cannot be explained on other grounds (nutrition, intoxication, deprivation, trauma, infection, autoimmunity, neoplasia, etc.)
3. When there is an unexplained or idiosyncratic response to a drug (i.e., onset of symptoms after phenobarbitol administration in acute intermittent porphyria, prolonged effect of succinylcholine given during anesthesia). This aspect of heritable disorders will be considered more fully in Chapter 14.
4. When the disorder at least partially involves changes in specific, measurable metabolites (e.g., urinary amino acids, blood ammonia, etc.) or when there are generalized metabolic aberrations (e.g., acidosis).

Thus, the possibility of an inborn error of metabolism should be considered when there is failure to thrive, a specific constellation of physical findings, abnormal results on routine or rational laboratory evaluation, deterioration after a period of normalcy, or unexplained adverse drug responses. In any of these circumstances, increasingly specific, rational screening tests,* including refined methods of physical examination (e.g., with a slit-lamp ophthalmoscope) or definitive tests (e.g., hemoglobin electrophoresis) can be utilized.

Specific clinical features that may suggest an inborn metabolic disorder include: a progressively downhill course once the disorder is manifested, with or without the development of organ (e.g., liver and spleen) enlargement; the development of unexplained renal or cardiac failure; changes in bone metabolism and growth; progressive loss of vision; coarsening of facial features; the development of kidney or bladder stones; unexplained anemia. This

*"Rational" tests are based on a logical assessment of the data in an individual case, and are to be distinguished from those employed uniformly, regardless of clinical circumstances (e.g., screening of all newborns for phenylketonuria) (8).

TABLE 6-1

A Brief Glossary of Terms Pertaining to Metabolic and Biochemical Disorders.

Amino acids. The building blocks or components of proteins; each is characterized by having at least one amine ($-NH_2$) and one carboxyl ($-COOH$) group.
Carrier test. Any test to determine whether a person is heterozygous for a recessive gene; usually involves sophisticated biochemical technology.
Chromatography. A means of separating metabolites using solvents without need for electrical gradients. For example, amino acids in urine can be separated from each other and identified very easily with this procedure.
Coenzyme (cofactor). A nonprotein component necessary for the activity of an enzyme; may be a vitamin.
Compound heterozygote. A cell or person with two different mutant alleles at a given locus; e.g., a person with hemoglobin SC disease has the mutant alleles S and C at that hemoglobin locus.
Deoxyribonucleic acid (DNA). The nucleic acid which makes up genes.
Electrophoresis. A process by which molecules are characterized and separated from each other, utilizing the different way each molecule travels under the influence of an electric (voltage) gradient.
Enzyme. A protein catalyst; it facilitates the conversion of one compound to another by adding, removing, or rearranging molecules. In living systems metabolic conversions depend on enzymes, and thus if an enzyme activity is deficient, the corresponding conversion of substrate to product doesn't take place effectively and metabolism is thereby disrupted.
Enzyme assay. A measurement of an enzyme's activity, determined by monitoring either the amount of substrate used up and/or the amount of reaction product formed under standard conditions. This method does not directly determine the amount of enzyme present, but, rather, the enzyme's conversion activity. With some mutations the amount of enzyme is normal but its activity has decreased or is absent under standard conditions.
Isoenzymes. Alternative forms of an enzyme catalyzing the same reaction. Different tissues may have slightly different forms of the same enzyme. For example, there are five forms of the enzyme lactic dehydrogenase (LDH).
Loading test. A means of determining how a person metabolically processes a standard amount (load) of administered substrate (e.g., glucose tolerance test). This can be an effective way of diagnosing affected persons or carriers of certain mutations.
Metabolite. A chemical substance that is part of a metabolic process, either as a substrate, or, more often, as a reaction product.
Polypeptide. A relatively short sequence of amino acids (e.g., 6–12); to be distinguished from a protein, which is longer and more complex.
Purine bases. Cyclic compounds belonging to one of the two classes of molecules (the other being pyrimidines) which are either the building blocks of nucleic acids (DNA, RNA) or their metabolites. They include adenine, guanine, hypoxanthine, xanthine, and uric acid.
Pyrimidine bases. Cyclic compounds belonging to one of the two classes of molecules (the other being purines) which are either the building blocks of nucleic acids (DNA, RNA) or their metabolites. They include thymine, cytosine, and uracil.
Reaction product. The substance or compound resulting from an enzyme's action on its substrate. The substrate is converted to the reaction product under the influence of the enzyme.
Regulator gene. A gene whose main or only function is to regulate or modulate the activity of a structural gene or genes.
Ribonucleic acid (RNA). A nucleic acid which is involved in various types of cell metabolism and especially the conversion of potential information in genes to actual gene products and gene expression.

Structural gene. A gene whose main or only function is to direct the synthesis of a polypeptide or a protein (in contrast to regulator gene).
Structural protein. A noncatalytic protein that functions in a structural or transport capacity (e.g., collagen and hemoglobin, respectively).
Substrate. The substance or compound on which an enzyme acts, which is converted to the enzyme's reaction product. Also, the substance or compound which is transported by a transport protein.
Transport protein. A noncatalytic protein that facilitates or causes the transport of a substance from one location to another. For example, ceruloplasmin, the protein deficient in Wilson disease, is the main transport protein for copper.

list emphasizes the progressiveness, diffuseness, and noncongenital character of most inborn errors of metabolism.

Heritable biochemical disorders are ultimately caused by a change in the structure of a gene that leads to the production of a functionally or structurally abnormal molecule, which in turn leads to cell or tissue dysfunction and thence to clinical problems (Table 6-1). Each cell's genes are located on chromosomes. Each gene corresponds to a certain length of a chromosome's deoxyribonucleic acid (DNA). The chromosomal DNA is closely associated with specific proteins, and together they are called chromatin. The DNA itself consists of two complementary strands of nucleotides. Each nucleotide is composed of a purine (guanine or adenine) or a pyrimidine (cytosine or thymine) combined in a specific way with a ribose sugar precisely attached to a phosphate group. If a segment of one strand of nucleotides is a purine, the corresponding segment of the complementary strand is a pyrimidine (actually, adenine is always paired with thymine, and cytosine with guanine). This pairing pattern defines why the DNA nucleotide sequences are described as complementary. The nucleotides are not sequenced randomly along the strand. Rather, each gene is known to contain serial groups of three nucleotides (triplets); that is, each gene consists of a specific series of nucleotide triplets. The first step in gene expression involves the synthesis of another nucleic acid—ribonucleic acid (RNA). The RNA is single stranded and it is synthesized at the gene site in such a way that its nucleotide triplets correspond exactly to one of the DNA strands, except that uracil replaces thymine. This "transcribed" RNA in turn serves as a mold, or template, from which a linear aggregation of amino acids can be synthesized (polypeptide or protein synthesis). Each amino acid of the protein is coded for by a specific triplet of RNA and of DNA in turn. The resulting protein directly reflects the structure of its gene; any change in the nucleotide sequence of the gene (mutation) changes the amino acid sequence, and therefore the structure and function, of the protein (9,10).

As stated above, the genetically altered protein molecule represents the primary defect in a single-gene heritable disorder (Fig. 6-1). This leads to changes in cell or tissue function at a level that depends on whether the protein serves as a structural element or a mediator of a specific biochemical process. The expression also depends on such considerations as whether the protein functions alone (monomer) or in combination with other proteins (polymer), and whether there are other mediators with overlapping or similar function (e.g., isoenzymes). Structural proteins contribute to cell structure (e.g., cytoplasmic and nuclear membranes) and to intercellular support (e.g., collagen). Biochemical mediators include enzymes and compounds which transport a substance from one place to another. Transport molecules can be either diffusible or membrane-bound. The immediate consequence of the primary defect is faulty structure of cells or tissues, inadequate transport of a substance within or between cells, or inadequate conversion of one compound to another. The chief intermediate consequences of enzyme and transport defects are the abnormal intracellular accumulation of substances (substrate) and/or intracellular deficiency of substances (product). In turn, the accumulations, deficiencies, or both, may have consequences that ultimately result in the clinical manifestation of the primary defect.

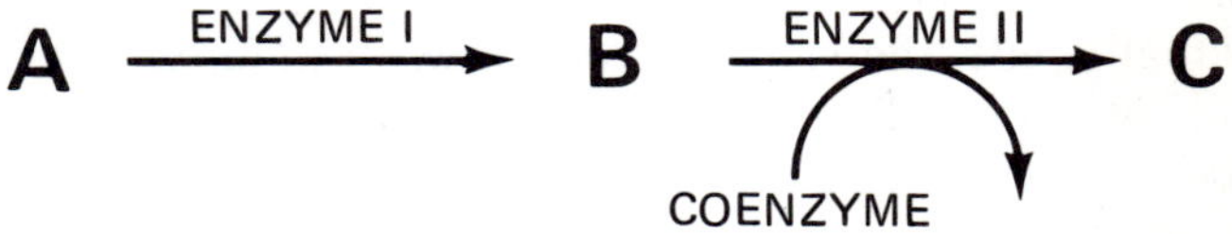

Block: Enzyme II concentration low.
To compensate:

If lack of C harmful

1. Add C directly
2. Add excess B to drive reaction to C
3. Add excess A to drive reaction to B to C
4. Add excess coenzyme to drive reaction to C
5. Supply enzyme II

If excess B harmful

1. Add excess coenzyme to drive reaction to C
2. Reduce concentration of A
3. Supply enzyme II

Figure 6–1 Simplified schema of possible levels of aberrant metabolism in biochemical disorders.

The presence of a mutant allele, or mutant alleles, may be inferred from the corresponding clinical disorder. The presence of the gene product (protein), mutant or normal, may be demonstrated in two basic ways:

1. By *measuring the function* for which the protein is responsible (i.e., its activity). For an enzyme, this means measuring how well the protein converts its substrate to its product under standard, defined conditions. For a transport protein, one usually measures how well the protein binds, or attaches to, the material to be transported. The activity of the mutant protein is then compared to the normal amount of activity and can be expressed as a percentage of the normal value. Reference to an enzyme deficiency usually means deficiency of the measurable activity.
2. By *measuring the amount* of protein actually synthesized, regardless of its activity. This can be done, for example, by seeing how much of an antibody against the protein is used up under standard assay conditions.

This distinction between deficiency of the protein and deficiency of its activity is important in several ways. In particular, it must be recognized that most statements of purported "enzyme deficiency" actually refer only to deficient activity: the gene product is merely altered, not absent.

General Genetic Principles

Since all mutations are said to entail a change in gene structure and a change in a protein's function, how do we relate such changes to a mutation's character as dominant, codominant, or recessive? When the primary defect involves an enzyme, the abnormality usually behaves clinically as a recessive (i.e., ordinarily no clinical problems in the heterozygote) and as a codominant in specific laboratory analyses (i.e., enzyme activity 50% of normal in heterozygotes, or electrophoresis of the enzyme shows both the normal and mutant enzyme patterns, etc.). This is also the case when the primary defect involves a diffusible transport protein (e.g., hemoglobin, ceruloplasmin). With membrane-associated transport or receptor proteins, the clinical expression is sometimes as a recessive (e.g., testicular feminization syndrome) and sometimes as a dominant (e.g., Albright hereditary osteodystrophy). Structural protein defects are usually manifested as clinical dominants, and there is rarely a direct laboratory assessment of the primary defect. Conversely, if a biochemical basis is known, a recessive pattern of inheritance almost always indicates an enzyme or diffusible transport defect, while a dominant pattern ordinarily suggests that the primary defect involves membrane-transport or a structural protein.

Tracing the primary defect in various inborn errors of metabolism has expanded our notion of the gene. It is no longer merely a functional conceptualization—the gene is now a biochemical entity which has properties that allow us to understand the cause and pathogenesis of many clinical disorders (11). This detailed knowledge also allows us to test various environmental agents, including drugs and medical x-rays, for their direct effects on genes (Chapter 7). In the long run, it may even become possible to alter heritable disease by actual manipulation of genes themselves (Chapter 15). This is a distant prospect, and efforts to attain the goal must be carefully monitored along the way.

Genetic heterogeneity (12) has become a useful clinical concept in biochemical genetics. What at one time seemed a single disorder has not infrequently turned out to be several, representing different mutations either at separate loci (nonallelic) or at the same locus (allelic). For example, there must be more than one gene locus for gout, since some affected individuals have a deficiency of a specific X-linked purine synthesis enzyme and others do not. There are many instances of allelic heterogeneity: dozens of additional mutations correspond to that for sickle hemoglobin (S), and there are dozens of mutations at the X-linked locus for the important glucose metabolizing enzyme G6PD (see Chapter 14). Moreover, these differences can be very subtle and yet have important management implications. Thus, hemoglobin D (Punjab), which is allelic to the normal hemoglobin A and mutant S, may have few or no clinical consequences, yet under ordinary assay (electrophoresis) conditions it is indistinguishable from the S form. By altering the laboratory conditions, the differences can, however, be demonstrated (13). This example demonstrates that heterogeneity is an important consideration when the specific diagnosis seems elusive or confusing, and because management may differ considerably from one form of the genetic disorder to another. This latter point is also evident in distinguishing the hyperphenylalaninemia of phenylketonuria (PKU) from benign disorders characterized by inconsequential elevations of serum phenylalanine (14).

Compound heterozygotes have two mutant alleles at a given locus. They are best appreciated in biochemical terms. Thus, a person with hemoglobin SC disease has two mutant alleles for the same locus, and this is readily apparent in the hemoglobin electrophoresis pattern, which shows both mutant hemoglobins instead of one or the other, as would be seen in a homozygote. (This is also an important consideration for the alpha-1-antitrypsin deficiency locus discussed in Chapter 4.) Compound heterozygosity should be considered in all autosomal recessive disorders where there is no parental consanguinity, especially if the clinical picture does not exactly fit the preliminary laboratory data. This is because unrelated parents may have distinct mutant alleles for disorder.

While demonstration of the primary biochemical defect is useful for both a basic and clinical understanding of pathogenesis, it must be emphasized that purely biochemical explanations also fall short in several ways. In particular, they do not alone account for all the variation in the respective disorders, both between and within families. This is especially true for dominant disorders where variable expression is characteristic. Other genetic as well as nongenetic factors, including specific and nonspecific environmental ones, must also be taken into account.

Clinical Evaluation of Inborn Metabolic Errors

General guidelines for suspecting a biochemical disorder were presented earlier in this chapter. Now we shall consider relatively specific indications and relatively specific tests (Tables 6-2 and 6-3). The intent is not to cover this area in exhaustive detail, but rather to give some idea of how one might think through a workup and what its value could be (Fig. 6-2). One basic principle must be clear from the outset: in every case, the most specific, definitive test available must be used before a given biochemical disorder is diagnosed with any certainty. This procedure should be obvious from the preceding section. There are two distinct reasons (elaborated in the next section) for being so specific: 1. definition of management, 2. definition of recurrence risks. The various types of tests employed will either demonstrate the primary defect (e.g., specific enzyme assay, hemoglobin electrophoresis) or assess its immediate consequences in terms of excessive production or accumulation of a metabolite, and/or deficient production or accumulation of other metabolites in various body tissues, fluids, and excretions. These assessments may be quantitative (specific measurements) or qualitative, as with fixed and stained pathology specimens or skeletal x-rays. Measurements of the defect can often be done by using intact cells and tissues, or extracts thereof. The actual tissue used depends on whether the defect in question shows itself there; for example, the enzyme deficiency of phenylketonuria can be directly demonstrated only in the liver, the tissue where the normal enzyme activity would ordinarily be found. Other tissues that can be used—depending on the disorder—include kidney, rectal mucosa, bone marrow, red blood cells, white blood cells, serum, amniotic fluid cells, and fibroblasts cultured from a skin biopsy. (Other tissues could also provide fibroblasts, but the skin is readily accessible and the biopsy a simple procedure.) The vast majority of definitive tests which require sampling of a tissue can be done on circulating red

TABLE 6-2
Evaluation for Inborn Errors of Metabolism—Frequently Used Screening Procedures and Common Indications

Alph-1-antitrypsin levels, Pi typing. suspected alpha-1-antitrypsin deficiency
Amino acid urine chromatography. suspected primary or secondary aminoaciduria
Amylase serum levels. suspected cystic fibrosis or hereditary pancreatitis
Berry spot test (urine). suspected mucopolysaccharidosis
Bilirubin serum levels (direct and indirect). suspected liver disease or hemolysis; evaluation of clinical jaundice
Calcitonin serum levels. suspected medullary thyroid carcinoma as part of the multiple endocrine adenoma syndrome II
Calcium serum and urine levels. suspected disturbance of calcium metabolism, e.g., in Albright hereditary osteodystrophy
Cholesterol serum levels. suspected primary or secondary hyperlipidemia; familial arteriosclerotic heart disease
Complete blood count. evaluation of all blood dyscrasias, primary or secondary
Copper urine levels. suspected Wilson disease or Menkes disease
Electrolytes. (Na^+, K^+, Cl^-, CO_2) serum and/or urine levels. Suspected or known disorders which lead to acidosis or alkalosis, and/or which involve hormones that act on the kidney (e.g., aldosterone, pitressin, etc.).
Galactosemia screen. suspected galactosemia
Glucose-6-phosphate dehydrogenase screen. semiquantitative assay in red blood cells of patients with hemolytic anemia or who are to be given certain drugs
Glucose serum and urine levels. (may include fasting, 2 hour postprandial, and/or glucose tolerance test specimens). suspected diabetes mellitus, secondary hyperglycemia, glycosuria, or hypoglycemia, etc.
Guthrie test. screening for phenylketonuria, etc.
Hemoglobin electrophoresis. suspected hemoglobinopathy (e.g., sickle-cell anemia, thalassemia); useful for detecting homozygotes and heterozygotes
Lactose tolerance test. suspected lactose intolerance (Other sugars can be used for other types of suspected intolerance.)
Organic acid urine levels. suspected organic aciduria (e.g., isovaleric acidemia, glutaric aciduria, etc.)
Osmotic fragility test. suspected hereditary spherocytosis
Porphyrin stool levels. suspected porphyria of all types
Porphyrin urine levels. suspected porphyria of all types
Steroid urine levels. suspected hypogonadism and/or inborn error of the adrenals or gonads
Sweat test. suspected cystic fibrosis
Tay-Sachs carrier screen. Ashkenazi Jewish descent
Triglyceride serum levels. suspected primary or secondary hyperlipidemia; familial arteriosclerotic heart disease
Uric acid serum levels. suspected gout or Lesch-Nyhan syndrome; followup of patients with psoriasis, Down syndrome, and glycogen storage disease I
Uric acid: creatinine ratio in urine. suspected gout or Lesch-Nyhan syndrome
Vanillyl mandilic acid urine levels. suspected pheochromocytoma (e.g., in multiple endocrine adenoma syndromes or neurofibromatosis) or neuroblastoma
Watson-Schwartz test. suspected acute intermittent porphyria
Zinc serum levels. suspected acrodermatitis enteropathica

or white blood cells or cultured skin fibroblasts. If a definitive diagnosis can be made without fibroblast culture, so much the better, considering the amount of work it takes to establish the culture and grow the required numbers of cells. Whenever an individual has a poorly understood or undiagnosed inborn error of metabolism, it is *always* worthwhile to prospectively arrange for postmortem assays on frozen organs obtained at autopsy shortly after death. Organs of special value are brain, liver, and kidney; lung, spleen, muscle, bone growth centers (epiphyses), or specific endocrine glands may be useful in certain instances. If the primary defect could be identified from these specimens, the inheritance pattern, the basis for a carrier test, or a treatment modality might become clear. Such studies could improve treatment of other affected children and allow the parents to more clearly understand the illness and recurrence risks.

Various body fluids can sometimes be used for diagnosis, including plasma, serum, and cerebrospinal fluid. So can excretions, including feces, urine, saliva, and tears. Urine, in particular, is very useful, especially when the disorder involves the kidney itself (i.e., many of the lysozomal storage diseases) or when there is a marked overproduction or decreased breakdown of certain metabolites, including amino acids, organic acids, and inorganic substances such as pyrophosphate and metallic ions (e.g., copper). Sometimes only a single sample is needed; in other cases a time-span (e.g., 24-hour) sample is needed. Occasionally, it is even necessary to do clearance studies, i.e., to trace the rate at which an administered dose of a substance is cleared, and see whether it has been metabolically converted and to what. Thus, there are many ways of reaching a specific biochemical diagnosis.

TABLE 6-3

A Battery of Six Urine Tests That Comprise a Fairly Complete "Metabolic Screen" in the Early Evaluation of a Patient Suspected of Having an Inborn Error of Metabolism

1. **Ferric chloride test.** Tests for the presence of phenylpyruvic acid or beta-imidazole pyruvic acid. Excess amounts of the former indicate phenylketonuria (PKU), and excess amounts of the latter indicate a deficiency of the enzyme histidine alpha-deaminase.
2. **Nitroprusside reaction.** A positive test is found if cystine is present, indicative of cystinuria.
3. **Dinitrophenylhydrazine test.** Positive test indicates the presence of ketoamino acids in the urine. A positive result is likely in PKU (phenylpyruvic acid), histidinemia (imidazole pyruvic acid), or methionine malabsorption. The test is positive with acetone (hyperglycinemia, isovaleric acidemia, and glycogen storage disease types I, III, V, and VI) and with ketonuria due to other causes.
4. **Benedict test for reducing substances.** Positive test when reducing substances are present in urine, including creatinine and uric acid. Abnormal reducing substances include a variety of sugars (glucose, lactose, fructose, galactose, maltose, mannose, and pentose) and homogentisic acid.
5. **Nitrosonaphthol test for tyrosine.** A positive test indicates the presence of tyrosine and its metabolites. The test will be positive with tyrosine transaminase deficiency, tyrosinosis (tyrosinemia), and transient tyrosinemia of the newborn.
6. **The Berry spot test for mucopolysaccharides.** A positive test indicates excess excretion of one or more of several mucopolysaccharides and indicates the presence of a mucopolysaccharidosis (e.g., Hurler syndrome, Morquio syndrome, etc.).

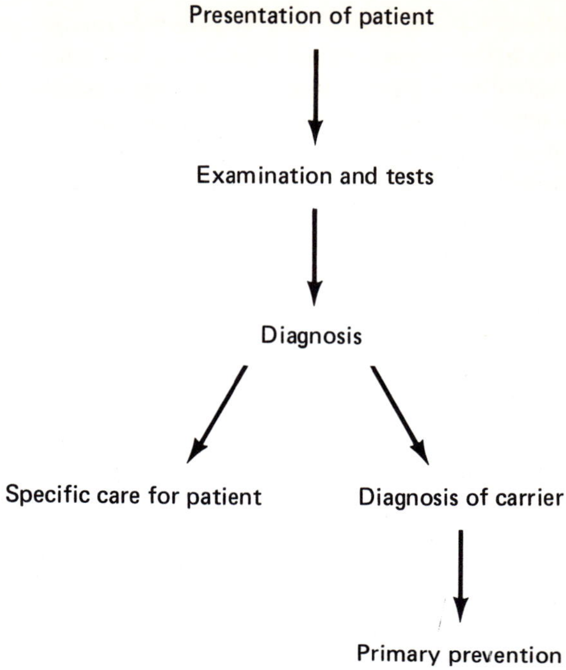

Figure 6–2 Schema of involvement with a patient and possible consequences of recognizing the basic biochemical lesion of an inborn error of metabolism. A specific diagnosis of a metabolic disorder can afford the possibility of primary, secondary, and tertiary prevention.

Often, however, single urine samples, 24-hour urine collections, single blood samples, or a skin biopsy are all that need to be considered.

Selected Disorders

Congenital biochemical disorders, apparent at birth, can potentially be diagnosed by the physical findings (Table 6-4). This is true of hypophosphatasia, the adrenogenital syndrome in a karyotypic female, and congenital hypothyroidism. *Hypophosphatasia* is manifested at birth by small body length; noncalcified cranium; severe, diffuse bone disease apparent on radiologic examination; low or absent serum alkaline phosphatase, and elevated 24-hour urinary excretion of phosphoethanolamine (15). Both parents should also be checked for heterozygote levels of blood phosphatase and urine phosphoethanolamine to document the autosomal recessive pattern since there is a variant dominant form. If the *adrenogenital syndrome* is suspected in a female newborn because of ambiguous genitalia, serum cortisol levels and 24-hour urine levels of 17-hydroxy- and 17-keto-steroids should be measured, serum electrolytes (especially sodium and potassium) monitored, and replacement therapy initiated before dehydration and shock develop (6). More complete endocrine studies to determine the level of the enzyme deficiency (e.g., 21-beta-hydroxylase deficiency) are important for long-term management. *Congenital hypothyroidism* may be sporadic or due to any of several heritable (autosomal recessive) thyroidogenesis deficiencies (16). Physical and x-ray findings usually suggest the diagnosis and various endocrine studies will delineate the lesion. Of course, a history of a previously affected sibling would facilitate diagnosis for each of these disorders.

A *neonatal* presentation usually involves a normal-appearing infant that develops acidosis or other metabolic disturbances within hours or days after birth. In a male infant with the *adrenogenital syndrome* (i.e., salt-losing type with 21-beta-hydroxylase deficiency), except for nipple and scrotal hyperpigmentation, the only indication might be hyponatremia leading to dehydration and shock. Diagnosis involves the approach described above, but the situation may be more urgent, requiring reliance on an indirect diagnosis through serum and urinary electrolytes. *Ornithine transcarbamylase deficiency* is an X-linked recessive defect in urea-cycle metabolism that shows up in the newborn nursery as acute metabolic acidosis with hyperammonemia (17). The specific diagnosis is further suggested by the pattern of excessive amino acids in the urine, but definitive diagnosis requires an assay performed on liver tissue obtained at biopsy or autopsy. Curtailment of dietary protein and other nonspecific therapies are appropriate. The *Zellweger syndrome* (cerebro-hepato-renal syndrome) usually appears within several days of birth as hypotonia, listlessness, and mild jaundice (18). Additional signs of impaired cerebral function and the development of gradual, progressive renal and hepatic failure ensue. This autosomal recessive disorder

TABLE 6-4
Metabolic Disorders That Are Relatively Common and/or for Which There Are Screening Tests Available

Adrenogenital syndrome
Alpha-1-antitrypsin deficiency
Cystic fibrosis
Diabetes mellitus
Galactosemia
Glutaric aciduria
Gout
Hereditary spherocytosis
Hurler syndrome
Hyperbilirubinemias
Hyperlipidemias
Lactose intolerance
Multiple endocrine adenomatoses
Phenylketonuria
Porphyrias
Sickle-cell anemia
Tay-Sachs disease
Thalassemias
Wilson disease

has an inexorable downhill course. The diagnosis is suspected from the combination of organ failures and is strengthened by finding elevated serum iron concentrations and pipecolic acid in the urine (19). Definitive diagnosis requires histologic examination of the liver and kidneys, including the use of iron stains.

Disorders presenting in *infancy* with the development of organ failure, especially the brain and liver, should always suggest an inborn metabolic error. For example, the onset of progressive jaundice and other signs of liver failure in an infant or child should raise the possibility of homozygous (or compound heterozygous) *alpha-1-antitrypsin (AAT) deficiency*. It may be suspected from a low or absent alpha-1 peak on routine serum protein electrophoresis, and documented by deficient levels of AAT activities in serum samples and by the presence of diastase-resistant PAS-positive inclusions in hepatocytes on liver biopsy. It is essential to determine the AAT (Pi) genotype of the proband and parents by immunoelectrophoresis for accurate prognosis and genetic counseling. Regression from attained developmental milestones and progressive deterioration, especially if combined with loss of vision, whether or not an enlarged liver and spleen and other visceromegaly develop, is a clear indication for an in-depth biochemical evaluation. Among the disorders to consider are all the *lipidoses* (e.g., Krabbe disease, Tay-Sachs disease, G_{m1}-gangliosidosis), *lysozomal storage diseases* (e.g., I-cell disease), and the mucolipidoses (e.g., fucosidosis, mannosidosis) (20–23). If spasticity develops, an *organic aciduria* (e.g., glutaric aciduria) should be considered (24). Evaluation for these disorders and their many variants includes a complete physical examination; a complete radiographic skeletal survey; urine samples (spot samples and 24-hour collections) for amino acids, organic acids, mucopolysaccharides, and lysosomal enzymes; a bone marrow sample to look for storage cells; serum and white blood cell samples for a battery of specific enzyme assays; and a skin biopsy for transmission electron microscopy (looking for particular types of cellular inclusions) and for culture of fibroblasts to be assayed also for a battery of enzymes. Even though treatment modalities may be inadequate and experimental, a reasonably rapid diagnosis can save the family a lot of heartache and unnecessary frustration, and it would also clarify recurrence risks and prognosis. If these techniques fail to provide an answer, then sampling of one of the target organs by biopsy or at autopsy is imperative for the reasons just noted.

In *childhood* these same types of disorders, and the *mucopolysaccharidoses* as well, may be present. The development of facial coarsening may be a useful sign. The evaluation would be done in the manner just described. Other disorders, such as *Fabry disease* (ceramide trihexosidase deficiency), usually take considerably longer to become manifest. The range for presentation with Fabry disease, an X-linked recessive trait, extends from childhood well into adulthood (25). The manner of presentation is extremely variable, and does not involve mental retardation: the disorder's features are cutaneous angiokeratomas over the trunk, groin, and genitalia; paresthesias; unexplained fever and sweating; heart failure; and renal failure which is a major cause of death. Enzyme assay of peripheral white blood cells or cultured skin fibroblasts as early as possible allows for specific diagnosis and thus close monitoring of renal function before failure sets in, permitting optimal timing of renal transplantation. The identification of female carriers is particularly important in this disease since they are at high risk for developing the renal complications.

Few inborn errors of metabolism present in the *adult* years, and of those that do, mental retardation is not among their features. Fabry disease has already been mentioned. Adult *Gaucher disease* (Type I; nonneuronopathic glucocerebrosidase deficiency) may have its onset in childhood, but adult presentation is frequent (26). Glucosyl ceramide accumulates in reticulum cells and cells with phagocytic abilities, especially endothelial cells, as a result of the glucocerebrosidase deficiency in organs such as the liver, spleen, and bone marrow. The resulting "Gaucher cells" are a hallmark of the disease. Clinical features include splenomegaly, thrombocytopenia, bone pain, and liver impairment without jaundice. Gradual encroachment on parenchyma by increasing numbers of Gaucher cells, especially in the spleen and bones, leads to the disabilities the disorder causes—even to death—although specific mechanisms are not always clear. This autosomal recessive disorder is suspected on clinical grounds (usually hypersplenism), and the suspicion is heightened by serum acid phosphatase elevations. A bone marrow specimen would show Gaucher cells. Definitive diagnosis of the enzyme deficiency can be made from circulating leucocytes or cultured skin fibroblasts. Both homozygous, affected patients and heterozygous, asymptomatic carriers can be diagnosed in this manner.

The heterozygous *carrier state* for many of the known enzyme deficiency disorders can be diagnosed in adults—either through a screening program (e.g., Tay-Sachs disease) or because of an affected relative. For a known carrier of an autosomal recessive trait, it would then be reasonable to also determine whether the spouse or intended spouse is a carrier. This approach is especially recommended for disorders with relatively high incidences in the general population (e.g., alpha-1-antitrypsin deficiency) or in specific ethnic groups, such as Tay-Sachs disease or Gaucher disease (Type I) in Ashkenazi Jews and sickle-cell anemia in blacks.

Management and Prevention

A recurrent theme in this chapter has been the practical consequences of demonstrating the primary defect in an affected person or carrier for a given disorder. These may be practical consequences for an individual patient and his family, or for relatively large populations. The individual family's main concerns relate to the affected person (definition of management) and to future members (definition of recurrence risks). The concern for large populations is the availability and efficiency of routine screening programs.

Definition of management. A specific biochemical diagnosis can establish the proper specific and nonspecific modes of therapy, and clarify the course of the disorder and ultimate prognosis. From the patient's and family's points of view each of these is a tremendously important reason for assuring the most accurate diagnosis. Even if there will be no specific therapy and the disease's course will be unchanged, there still may be reasons for clarifying the diagnosis, including an end to an elusive search for the cause.

Definition of recurrence risks. If the primary defect involves an enzyme or diffusible transport protein, the mutation is most likely a recessive. There are only two consistent, well-documented autosomal dominant enzyme defects, and it is not even clear that they represent the true primary defects [uroporphyrinogen I synthetase deficiency in acute intermittent porphyria (27) and phosphoribosylpyrophosphate synthetase excess with a certain form of gout (28)]. If an enzyme deficiency affects a female, it is probably an autosomal defect. If there is only an affected male, then the enzyme deficiency might be shown to be autosomal by previous experience (i.e., known autosomal locus) or by demonstrating that both parents are carriers. If only the mother is a carrier, and paternity is not in doubt, then an X-linked locus is likely. Once the mode of inheritance is clear and carriers are identified if possible, then meaningful genetic counseling can be provided.

Screening programs for biochemical disorders depend on knowledge of the primary defect, the burden of the disorder, the cost of the screening, and the sensitivity and specificity of the screening test. It is important to distinguish between programs designed to detect affected persons (e.g., phenylketonuria, galactosemia), carriers alone (e.g., Tay-Sachs disease carriers), or both (e.g., sickle-cell anemia, alpha-1-antitrypsin deficiency). In each case genetic counseling must be provided to families and persons having positive results: It is wrong to demonstrate the presence of a mutant gene without clarifying its significance. Additional ethical and legal aspects of screening are discussed in Chapter 15.

Inborn errors of metabolism have a special place in modern

medicine. In terms of individual families, they epitomize the rational approach to diagnosis, treatment, and prevention of future occurrences. In terms of large populations, they hold the promise of minimizing the proportion of children who would suffer disabling mental retardation or early death. This is the portent of screening programs: to identify the affected youngster and begin treatment before there are irreversible changes, and to identify carriers who could then adopt any of several alternatives to avoid having an affected child. Also, in terms of large populations, specific biochemical disorders seem to hold the key to understanding the mechanism of more common disorders such as cancer, arteriosclerotic heart disease, hypertension, and diabetes mellitus. Already, for example, elucidation of the primary defect for an uncommon form of hypercholesterolemia (29) raises the prospect of ultimately decreasing the prevalence and morbidity of atherosclerotic heart disease. Nonetheless, biochemical disorders are often considered to have little significance in day-to-day clinical practice. An effort has been made in this chapter to counter this notion by clarifying the fact that, in the aggregate, these disorders are not so rare; they can be appreciated largely in terms of such general concepts as age of onset and progressive deterioration after a period of normalcy. Recognition and treatment are facilitated by the need to examine only a relatively small number of body fluids, excretions, and tissues; and by the availability of clinical and biochemical geneticists to assist in the evaluation.

Cited References

1. Pearson HA, Diamond LK: Sickle cell disease. *Pediatrics* 48:629, 1976.

2. Seegmiller JE: Inherited deficiency of hypoxanthine-guanine phosphoribosyl transferase in X-linked uric acidura (the Lesch-Nyhan syndrome and its variants). *Adv Hum Genet* 6:75, 1975.

3. de Brayn CHMM: Hypoxanthine-guanine phosphoribosyl transferase deficiency. *Hum Genet* 31:127, 1976.

4. Hsia DY-Y: Phenylketonuria and its variants. *Prog Med Genet* 7:29, 1970.

5. Wong PWK, Justice P, Smith GF, et al: A case of classical maple syrup urine disease, 'thiamine non-responsive'. *Clin Genet* 3:27, 1972.

6. Bongiovanni AM: Disorders of adrenocortical steroid biogenesis. in Stanbury JB, Wyngaarden JB, Fredrickson DS (eds): *The Metabolic Basis of Inherited Disease,* (3rd ed), New York, McGraw-Hill, 1972, pp 857–885.

7. Danks DM, Campbell PE, Stevens AJ, et al: Menke's kinky hair syndrome. An inherited defect in copper absorption with widespread effects. *Pediatrics* 50:188, 1972.

8. Smith DW, Simons FER: Rational diagnostic evaluation of the child with mental deficiency. *Am J Dis Child* 129:1285, 1975.

9. Stanbury JB, Wyngaarden JB, Fredrickson DS: Inherited variation and metabolic abnormality. in Stanbury JB, Wyngaarden JB, Fredrickson DS (eds): *The*

Metabolic Basis of Inherited Disease (3rd ed), New York, McGraw-Hill, 1972, pp 3–28.

10. Hoagland MB: Coding, information transfer and protein synthesis. Ibid, pp 29–51.

11. Harris H: *The Principles of Human Biochemical Genetics.* Amsterdam, North Holland, 1975, pp 1–29.

12. Childs B, Der Kaloustian VM: Genetic heterogeneity. *New Engl J Med* 279:1205, 1968.

13. Stout C, Holland CK, Bird RM: Hemoglobin D in an Oklahoma family. *Arch Intern Med* 114:296, 1964.

14. Blaskovics ME, Schaeffler GE, Hack S: Phenylalaninemia: Differential diagnosis. *Arch Dis Child* 49:835, 1974.

15. Teree TM, Klein L: Hypophosphatasia: Clinical and metabolic studies. *J Pediatr* 72:41, 1968.

16. Fraser GR: The genetics of thyroid disease. *Progr Med Genet* 6:89, 1976.

17. Campbell AGM, Rosenberg LE, Snodgrass PJ, et al: Ornithine transcarbamylase deficiency: A cause of lethal hyperammonemia in males. *New Engl J Med* 288:1, 1973.

18. Passarge E, McAdams AF: Cerebro-hepato-renal syndrome. *J Pediatr* 71:691, 1967.

19. Danks DM, Tippett P, Adams C, et al: Cerebro-hepato-renal syndrom of Zellweger. *J Pediatr* 86:382, 1975.

20. Brady RO: The lipid storage diseases: New concepts and control. *Ann Intern Med* 83:257, 1975.

21. Dekaban AS, Herman MM: Childhood juvenile and adult cerebral lipidoses. *Arch Pathol* 97:65, 1974.

22. Desnick, RJ, Thorpe SR, Fiddler MR: Toward enzyme therapy for lysosomal storage disease. *Physiol Reviews* 56:57–99, 1976.

23. Kelly TE: The mucopolysaccharidoses and mucolipidoses. *Clin Orthop Rel Res* 114:116, 1976.

24. Watts RWE, Chalmers RA, Lawson AM: Abnormal organic acidurias in mentally retarded patients. *Lancet* 1:368, 1975.

25. Clarke JT, Knaack J, Crawhall JC, et al: Ceramide trihexosidosis (Fabry's disease) without skin lesions. *New Engl J Med* 284:233, 1971.

26. Ho MW, Seck J, Schmidt D, et al: Adult-Gaucher's disease: Kindred studies and demonstration of a deficiency of acid beta-glucosidase in cultured fibroblasts. *Am J Hum Genet* 24:37, 1972.

27. Tschudy DP, Valsamis M, Magnussen CR: Acute intermittent porphyria: clinical and selected research aspects. *Ann Intern Med* 83:851, 1975.

28. Becker MA, Meyer LJ, Wood AW, et al: Purine overproduction in man associated with increased phosphoribosylpyrophosphate synthetase activity. *Science* 179:1123, 1973.

29. Goldstein JL: Genetic aspects of hyperlipidemia in coronary heart disease. *Hosp Pract* 10:53, 1975.

General References

A. Ampola MG, (ed): *Early Detection and Management of Inborn Errors.* Clinics in Perinatology, vol 3, No 1, Philadelphia, WB Saunders, 1976.

B. Brady RO: Inherited metabolic diseases of the nervous system: *Science* 193:733, 1976.

C. Conley CL, Charashe S: Inherited hemoglobinopathies. in McKusick VA, Claiborne R (eds): *Medical Genetics*, New York, HP Publishing, 1973, pp 53–61.

D. McKusick VA: *Heritable Disorders of Connective Tissue* (4th ed) St. Louis, CV Mosby, 1972.

E. Scriver L. Rosenberg L: *Amino Acid Metabolism and its Disorders*. Philadelphia, WB Saunders, 1973.

F. Stanbury JB, Wyngaarden, JB, Fredrickson DS (eds): *The Metabolic Basis of Inherited Disease*. New York, McGraw Hill, 1972.

CHAPTER SEVEN

GENETIC AND CONGENITAL DISORDERS—THE ADVERSE EFFECTS OF ENVIRONMENTAL AGENTS

Introduction

The environment influences genetic expression in all phases of life, including fetal development (Fig. 7-1). Although there is always interaction between these elements, one or the other may predominate at certain times (Fig. 7-2). Optimal environmental circumstances, for example, will not overcome the adverse effects of an extra chromosome 21 in the Down syndrome. On the other hand, a genetically normal child born to a mother with untreated phenylketonuria will have a low birthweight, microcephaly, and—ultimately—severe mental retardation. Underlying other diseases there is a mutual interplay of these two sets of determinants, as in the various polygenic disorders (e.g., neural tube defects, diabetes mellitus). A clinical geneticist's task, for any given disorder, is to determine the relative importance of each factor. After that, the recurrence risks and basis for prevention can be determined in a given family. This chapter is intended to outline general principles governing the effects of environmental agents on genetic material and its expression; to describe the adverse effects of clinically important environmental agents such as x-rays, progestational agents, and rubella virus; and finally to demonstrate how these considerations influence patient care.

Health care workers are often asked about the adverse effects of environmental agents, particularly in regard to a previous, current, or future pregnancy. The present concern of the public and government agencies with pollutants, pesticides, food and drug safety, and informed consent has largely shifted the focus from retrospective to prospective. In clinical practice, however, both perspectives are important: the *retrospective* to *account* for an

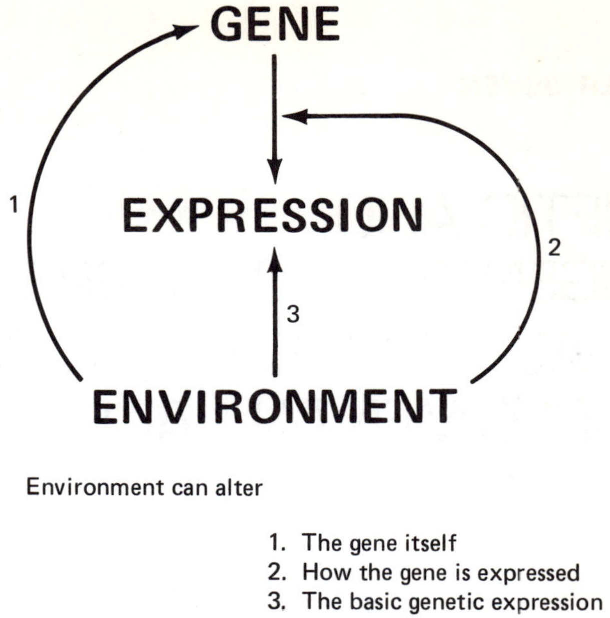

Figure 7–1 Interplay of genes and the environment in the expression of a genetic trait. Genes are expressed not only in the context of, but also under the influence of, environmental factors.

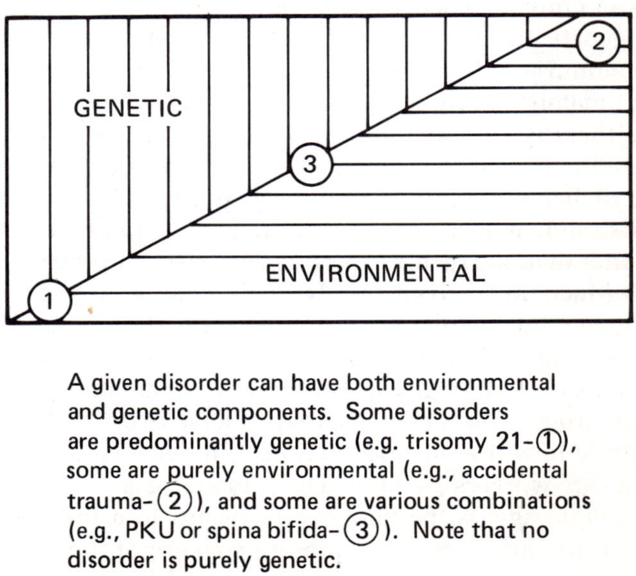

Figure 7–2 Relative contributions of genetic and environmental factors in the causation of disease.

existing problem, and the *prospective* to *prevent* or minimize the likelihood of occurrence. Thus, some inquiries about the adverse effect of an agent are prompted by the birth of a child with a congenital or genetic disorder (retrospective). Other inquiries are prompted by either an anticipated pregnancy or a current pregnancy in an early stage (prospective). In genetic counseling, the two are sometimes combined; a couple has had a congenitally abnormal child and there is concern about recurrence of the disorder in a pregnancy planned for the near future. If the retrospective assignment of an environmental cause can be made, the prospective recurrence risks can be defined. If the environmental agent can be avoided or will not be a factor (e.g., syphilis, maternal rubella), then there is no substantial prospective risk. If the environmental agent cannot be avoided (i.e., specific therapy with diphenylhydantoin or anticancer agents), then the prospective risk may be substantial.

It must be stressed, however, that, except for a relatively small number of environmental agents, a specific, quantifiable, adverse effect cannot be ascribed, prospectively or retrospectively. Thus, it often becomes necessary to balance the general admonition against unnecessary exposure to any potentially harmful agents with the apparent necessity of exposure to them. Every specific exposure occurs in the context of a general background risk. Thus, for example, if a pregnant woman is encouraged to learn in counseling that her exposure to irradiation from a chest x-ray is of little concern, she should also understand that there is no guarantee that her baby will be free of defects, and that the risk for major malformations in the general population is about 3% (Chapter 2).

Table 7-1 lists the general categories of environmental agents and their possible adverse effects on genetic expression and embryologic development. It is important to recognize that each type of agent can have more than one adverse effect and that each adverse effect can be caused by more than one type of agent. The rest of this chapter will deal with these relationships in terms of clinical problems.

The Adverse Effects

Of the five effects listed in Table 7-1, at least three directly affect the genetic material at various levels of organization (mutagenesis, oncogenesis, and chromosome changes) while for the other two (teratogenesis and very early abortion) it is not possible to rule out an agent's interference at another biochemical level (as a general or specific toxin, for instance). It is important to appreciate these distinctions, since the first three may be both clinically inapparent and transmissible. Thus, an apparently normal infant exposed to any of various agents in utero may still be at risk for

TABLE 7-1
Environmental Agents and Their Potential Adverse Affects

- I. Agents
  - A. Irradiation
  - B. Infectious
  - C. Pharmacologic (chemical)
  - D. Maternal
- II. Adverse affects
  - A. Mutagenesis
  - B. Oncogenesis
  - C. Chromosome changes
  - D. Teratogenesis
  - E. Very early abortion or resorption

adverse consequences either in himself (e.g., cancer) or in his offspring (e.g., mutations). An individual with a teratogenic malformation usually is not considered at risk for transmitting the defect itself to offspring, but germinal mutations might be present as an independent effect of the teratogen. Although immediate assistance can now be given to affected families through knowledge of the teratogenic potential of various agents, particularly irradiation and certain drugs, in the long run, perhaps, it is direct changes in the genetic material itself that will be of greatest importance.

Mutagenesis refers to a change in a gene that gives it a new biochemical structure and a new functional expression. The vast majority of mutations, spontaneous or induced, are deleterious. Most, however, are recessive, and thus are inapparent in the heterozygote and may not be expressed clinically for many generations (until the mutant allele is paired with another mutant allele at the same locus). Clinically dominant mutations are expressed as they occur. If the dominant mutation is germinal (occurring in a parental egg or sperm or in the zygote), then the individual not only is at risk of expressing the mutant, but also of transmitting it in accordance with the principles of dominant inheritance (Chapter 4). If the dominant mutation is somatic (occurring in a postzygotic nongerm cell), it will be expressed only in the tissues derived from the cell in which the change took place. [Whether somatic mutations may involve a two-step process is a fascinating consideration, but beyond the scope of this text. Interested readers are referred to Knudson et al. (1).] Dominant somatic mutations may be manifested functionally (e.g., hormone overproduction), or morphologically as congenital malformations (e.g., the prune-belly syndrome) or malignant neoplasia (e.g., retinoblastoma).

The relationship between mutagenesis and the other adverse effects is not always clear. Although some congenital malformations and malignancies can be explained in terms of mutation, teratogenesis is not ordinarily considered a mutational phenomenon (2), nor is oncogenesis to be equated with mutagenesis.

Chromosome breakage may involve mutations, but their presence is hard to assess because so much genetic material is involved or because the gene in question may merely have been deleted.

Oncogenesis involves a change in the growth and metabolic behavior of cells: they are no longer responsive to the ordinary restriction on proliferation. There seem to be many things that can bring about the "transformation" of a normal cell to a neoplastic cell. The reasons for assuming that genetic mechanisms are involved derive from at least three lines of evidence:

1. Daughter cells derived from a malignant cell maintain their malignant properties through successive cell generations
2. Many instances of cancer development are either preceded by or associated with chromosomal changes (e.g., the Philadelphia chromosome)
3. Many oncogenic agents are known to be mutagenic as well (3).

One of the problems in assessing the oncogenic nature of a given agent is the time lag between exposure to the agent and the development of cancer. This lag usually involves years. Very different types of agents are known to be oncogenic, but they apparently have in common the ability to somehow alter, alone or with other factors, the cell's genetic regulation of growth and proliferation. It is well known that some agents used to treat cancer are also mutagenic and/or oncogenic themselves (4).

Chromosome changes may be alterations in the number or structure of chromosomes. Environmental influences on the development of changes in chromosome number (aneuploidy) are just now being appreciated (5). Changes in chromosome structure can be produced by a variety of agents and a variety of mechanisms. Several heritable disorders (e.g., Bloom syndrome, Fanconi anemia, and ataxia telangiectasia) are characteristically associated with a propensity for chromosomal breakage and rearrangement of a fairly marked, easily measurable degree. It is quite possible that there may also be more subtle inherited tendencies in this direction, so that some persons are more susceptible than others to the clastogenic (chromosome-breaking) effect of environmental agents. In addition, breaks at some chromosome sites may have special importance, including the break point on the long arm of chromosome 22 characteristic of chronic myelogenous leukemia.

Teratogenesis involves structural aberrations in embryologic development for which environmental causes alone can be specified. (When no cause, genetic or nongenetic, can be specified, we merely refer to developmental anomalies). Teratogenic mechanisms vary according to the agent and target organ or tissue. To ascribe a teratogenic effect to a given agent, one should be able to show a consistent pattern of aberrations when fetuses are exposed to the agent at the same critical stage of development. A teratogenic effect is a potential consequence of all environmental agents, suggesting that without an explicit and sound reason, the fetus should not be exposed.

A recurrent difficulty in determining whether agents may be teratogenic in humans relates to the use of animal models as test systems. Teratogenic susceptibility varies from species to species and from one strain to another. The fact that an agent causes a defect in a strain of mice does not mean that it will have that effect in humans. Conversely, the absence of animal teratogenesis does not eliminate the risk for human fetuses. For more detail on the principles and methods of teratology, the reader is referred elsewhere (6).

Very early abortion (resorption) occurs so early that the conception is manifested only as a missed menstrual period or an unusually heavy menstrual flow, or by no signs at all. Thus, it is very difficult, if not impossible, to document and quantify among women. Still, environmental agents are presumed to cause very early abortion because this is a general phenomenon that occurs in a variety of animal species (7,8). It could be produced by any of the four mechanisms discussed above, or by a toxic metabolic effect.

The Environmental Agents

Each of the four classes of agents can have multiple effects on genetic material and its expression. Here we will consider these agents only in terms of the adverse consequences discussed in the preceding section, ignoring such things as nonneoplastic somatic tissue damage. The agents' adverse effects in the form of mutagenesis, oncogenesis, chromosome changes, and very early abortions will be discussed only briefly. Their teratogenic effects will be described in more detail.

Radiation has several sources and is of several types (Table 7-2). *Ultraviolet radiation,* either from the sun (actinic) or artificial sources, can damage DNA directly and thereby lead to somatic mutations and oncogenesis in exposed parts, primarily the skin. The mutations *per se* are probably of little consequence except as they contribute to oncogenesis. The ability of actinic ultraviolet radiation to cause skin cancer in farmers and sailors is well known. This adverse effect is tremendously magnified in the autosomal recessive disorder xeroderma pigmentosum, wherein the primary defect is deficiency of the enzyme which repairs ultraviolet-damaged DNA (9).

The adverse effects of *ionizing radiation* are an important modern concern. Sources include "background" atmospheric radiation; man-made nonmedical instruments such as atomic bombs, nuclear reactors, and radioactive mining equipment; and medical

TABLE 7-2
Common or Otherwise Important Sources of Potentially Damaging Irradiation

1. Excessive or improperly timed diagnostic x-rays.
2. Nuclear reactions, including excessive fallout.
3. Therapeutic irradiation.
4. Normal background irradiation for an overly sensitive individual.
5. Occupational exposure—x-ray machines, uranium mines, radar equipment, etc.

exposure for both diagnosis and therapy. Among the forms of medical irradiation are x-rays and radioisotopes used for diagnosis and treatment [e.g., radioactive iodine (I^{131}) in the evaluation and therapy of thyroid disorders].

Regardless of the source, the effects of ionizing irradiation are cumulative. Because it is difficult at best to quantify the effects of the background exposure, it is not clear whether this source contributes significantly to human medical problems of any type. Radiation from any of the other sources is more likely to be harmful, and the degree of concern is proportionate to the intensity and duration of exposure. It is not always clear, however, when a given level of exposure warrants special measures or precautions. For example, does a pregnant x-ray technician of ten years' experience and no documented "excessive" exposure have an increased risk for her fetus? What if she has had fifteen years of experience? And what if she herself has a polygenic birth defect (e.g., cleft palate)? Sooner or later we reach situations where the risk is clearly increased, but this does not afford a good answer for the in-between cases. In the vast majority, counseling regarding the adverse effect of irradiation relies mainly on general, rather than specific, guidelines.

Nonetheless, an effort should be made to assess the amount of radiation exposure and the proportion available to the fetus, gonads, or other target organs. There is no direct way to determine whether the irradiation of ordinary diagnostic procedures will have a significant adverse effect. For a pregnant woman, that uncertainty may be a factor in deciding to terminate the pregnancy.

Medical sources of radiation are important because they are used commonly and the dose and timing are known precisely. Not infrequently, x-rays are employed during pregnancy for legitimate reasons or inadvertently. To minimize the latter, some institutions require that a woman's last menstrual period be noted on x-ray requisitions and they will perform routine or elective examinations only during the first ten days since the period began. In addition to diagnostic x-rays, there are various types of diagnostic radioisotope scans, e.g., thyroid, liver, pancreas, bone, brain, kidney. Other tests may use isotopes without scans, as in the chromium^{51} red blood cell survival measurement. Therapeutic medical irradition includes x-rays, gamma rays, proton beams, and radioisotopes (e.g., radium, I^{131}). Therapeutic doses certainly must be avoided during pregnancies, especially in the early stages.

If a nonpregnant person has been diffusely exposed to radiation, there is no way to evaluate what the specific genetic consequences might be. Chromosome analysis of circulating leucocytes or other cells would only indicate whether or not there was chromosomal damage. Positive results could not forecast the risk for the development of cancer or for genetic disease in offspring. Thus, counseling would have to be very general. If a nonpregnant person received high-dosage irradiation in the area of the gonads, the danger of adverse genetic effects would be much greater but still not predictable. Men so exposed may develop transient or even permanent oligospermia or aspermia. High-dosage irradiation of the abdomen of a pregnant woman is certainly dangerous, but there is little direct evidence other than the Nagasaki and Hiroshima experiences (10). In none of these instances would amniocentesis for fetal chromosome analysis be in order, since there would be no way to interpret a negative result (normal chromosomes). That is, decisions about pregnancy should be made without attempts at pre- and postconceptional chromosome monitoring.

Among diagnostic and therapeutic isotopes, radioactive iodine is especially important. It is frequently used; it is intensively concentrated in the thyroid gland from about the thirteenth week of gestation on; and it is considered to be teratogenic, at least in therapeutic doses (11). Only for specific and compelling reasons should radioactive isotopes be administered to a woman who is known to be pregnant, and, even then, only with her full appreciation of the potential risk to the fetus.

Infectious agents, especially viruses, are important to the geneticist from several standpoints. The nonviral agents of concern include *Treponema pallidum* and *Toxoplasma gondii* for their causation of congenital syphilis and congenital toxoplasmosis, respectively. Viruses are known to cause several specific congenital syndromes and are tentatively implicated in other, less specific ones; this will be discussed in depth below. In addition, some viruses are known to cause transient chromosome breakage (12), so that recent "viral illnesses" should be considered in the interpretation of chromosome analyses. Viruses of various types also seem increasingly to account for at least some forms of cancer, including Burkitt lymphoma (13). If so, they might then explain some instances of familial cancer. The possibility of inherited susceptibility would still remain, but the emphasis in delineating recurrence risks would shift. The role of viruses and other infectious agents in mutagenesis is not known with any certainty at all.

Chemicals and *pharmacologic agents* are important causes of the adverse reactions listed in Table 7-1 and of the modifications in the expression of genetic disorders to be discussed in detail in Chapter 14. Well-known teratogenic effects will be considered below. Mutagenic effects are difficult to establish for pollutant chemicals from industrial sources. Mutagenic effects of certain drugs are well established, particularly, those of the anticancer drugs characterized as radiomimetic. Mutagenic drugs are also very often oncogenic, but the converse is not clearly the case. A number of hormonal drugs, for example, are at least potentially oncogenic, though not apparently mutagenic: conjugated estrogens from pregnant mare urine are correlated with endometrial cancer (14,15), birth control pills with hepatomas (16), and diethylstilbestrol with vaginal cancer in girls whose mothers consumed the drug during their pregnancy (17). At least two industrial pollutants are known to contribute to the development of malignant neoplasms: asbestos and lung carcinoma or mesothelioma, and vinyl chloride and hepatoma. Changes in chromosomes, especially breakage (clastogenesis), have been claimed to result from many chemicals and drugs, though not always convincingly in terms of exposure in intact humans. For example, caffeine in large amounts is associated with clastogenesis in tissue culture cells, but there is no similar effect with maximal coffee drinking (18). It is noteworthy that *none* of the illicit drugs, including marijuana, lysergic acid diethylamide (LSD), and heroin, have been established as clinical chromosome breakers, although there were claims to the contrary (19).

Maternal effects involve either the transmission of infections associated with birth or teratogenesis. The latter will be dealt with below. The important birth-associated infections result from passage through the birth canal [gonorrhea (ophthalmia neonatorum), cytomegalovirus, Herpes virus] or from maternal–fetal blood admixture (viral hepatitis).

Irradiation Teratogenesis

It is unusual for the developing fetus to be exposed to large amounts of radiation. When this does happen, the resulting abnormalities include microcephaly, skull malformations, other skeletal malformations, and general failure to thrive. With heavy irradiation there is rarely a dilemma as to expectations and reasonable courses of action, since the potentially adverse consequences are so obvious. With lesser amounts the situation is murkier, and decisions have to be made on general principles. Occasionally, for example, a woman with nonspecific gastrointestinal and urinary tract symptoms (e.g., anorexia, nausea, belching, flatulence, urinary frequency), undergoes an oral cholecystogram, upper gastrointestinal series, barium enema, and intravenous pyleogram, and with the final review of the x-rays, pregnancy is recognized. Given good x-ray equipment and standard x-ray doses, there is no basis for inordinate concern, *but* neither could the pregnant woman be told that there is no risk to the fetus at all. The key here is to introduce other factors into the equation. If the patient were a young woman with an unplanned first pregnancy and no difficulty conceiving, then even a very small risk might seem excessive since there is the alternative of abortion and starting again. Prospective counseling about diagnostic x-ray exposure often takes into account other things besides the exposure itself.

On the other hand, there is a more definite risk for a woman with flagrant hyperthyroidism, treated with I^{131}, who later finds that she was 13–14 weeks pregnant at the time of therapy. Teratogenesis in these circumstances has been well documented (11). Thirteen weeks may be critical since at about that time the fetal thyroid begins to concentrate iodine. Smaller exposures, as for diagnostic studies (thyroid scan and uptake), and earlier exposures are of less clear significance.

Thus, in summary, irradiation can be teratogenic, though the occasions where there is a clear potential for abnormalities to develop are unusual: nuclear reactors, maternal cancer, maternal hyperthyroidism, etc. There is no way to implicate paternal radiation exposure in teratogenesis, except through mechanisms of mutagenesis and chromosome changes.

Infectious Teratogenesis

Although only a few infectious agents are documented teratogens, others are strongly suspected and still others are presumed to be capable of teratogenesis, at least by analogy. This discussion will be confined to agents associated with more or less specific syndromes (Table 7-3). The overriding principle is that for a multiply malformed infant, an infectious agent must be sought, especially if the brain or eyes are involved and if there is no other specific diagnosis and the chromosomes are normal. With rare exceptions, the demonstration of an intrauterine infection as the cause of the abnormalities bodes well for subsequent pregnancies. The diagnostic evaluation includes a careful gestational history, physical examination of mother and child, serologic tests in both, cytologic evaluation, cultures for the offending agent in both, and if the child dies, an autopsy. Elevated IgM levels in cord and infant sera are at best a nonspecific indication of intrauterine infection. This finding must be followed by more specific studies. In this regard a TORCH screen is particularly helpful: the infant serum is tested for antibodies against *TO*xoplasmosis, *R*ubella, *C*ytomegalovirus, and *H*erpes simplex II. Positive results have

TABLE 7-3
Infectious Agents Definitely Known to Be Teratogenic for Humans

1. *Treponema pallidum* (syphilis)
2. *Toxoplasma gondii*
3. Rubella virus
4. Cytomegalovirus
5. Herpes simplex II virus

public health significance, since some of the viruses are excreted for months or years after birth.

Congenital syphilis is unique in that infection at any time during pregnancy can have adverse effects, though the severity is greater with early exposure (20). The diffuse manifestations represent infectious activity at each site. Stillbirths account for 70% of infections. In those who survive there are many immediate and long-range problems. The classical features include snuffles (mucopurulent nasal discharge which excoriates the upper lip), maculopustular cutaneous eruptions, scaling of the palms and soles, mucocutaneous fissures, keratitis, anemia, osteochondritis, perichondritis, hepatosplenomegaly and neurosyphilis in various forms. Prompt antibiotic treatment may help minimize the long-term consequences. Positive serologic findings in mother and child, plus demonstration of the spirochete in various lesions establish the diagnosis. Maternal syphilis can develop during subsequent pregnancies from reinfection, so that recurrence among future offspring is possible unless maternal disease is prevented.

Congenital toxoplasmosis usually results from a silent primary infection of the mother, although at times the maternal infection may be clearly manifested (21,22). The pregnancy may end in a stillbirth, or a premature or term infant. The common stigmata include chorioretinitis, cerebral calcifications, psychomotor retardation, hydrocephalus, and microcephaly. The diagnosis should be considered in any infant that does not progress as expected, especially if there is hydrocephalus or microcephaly. A good ocular fundus examination, under general anesthesia if necessary, skull x-rays, and examination of maternal and fetal sera for anti-toxoplasmosis antibodies help make the diagnosis. Demonstration of the parasite *Toxoplasma gondii* on stained smears from involved tissues or in excretions, of course, establishes the diagnosis. There is no particular concern for recurrence in subsequent pregnancies. The source of the parasite is not usually clear, though the domestic cat is thought to be a reservoir, excreting the protozoan in its stool. Women who are pregnant or planning to have a child thus are usually advised to have other household members clean up after the pet cat.

The *rubella syndrome* develops in a sizable proportion of pregnancies in which a woman contracts a primary infection with the rubella (German measles) virus in the first trimester (23). As vaccination against rubella may be a source of the virus, women who are pregnant or likely to become pregnant should not be vaccinated. The characteristic features of the rubella syndrome are small size for gestational age, microcephaly, cataracts, patent ductus arteriosus, deafness, and mental retardation. These features can vary considerably and may be limited to deafness that is not even recognized until the child is a year or more old. Moreover, the maternal primary infection may have been so subtle as to go unrecognized until the birth of an affected child. Knowledge that the mother had antirubella antibodies is helpful in ruling out the diagnosis. It is not yet known whether young girls (prepubescent) vaccinated against rubella will also be protected from having a child with the rubella syndrome. Naturally acquired immunity does apparently endow this protection. Evaluation of an infant for rubella must take place before the routine vaccinations at about one year of age; after that, the presence of antibodies may only reflect the vaccination.

Congenital cytomegalic inclusion disease (CID) should always be considered when a premature or term infant has signs of central nervous system dysfunction (hypotonia, developmental delay, blindness, deafness), especially if there is microcephaly or intracranial calcifications, hepatosplenomegaly, jaundice, anemia, and thrombocytopenia (24,25). A specific diagnosis must be confirmed as soon after birth as possible for at least two reasons: 1. The cytomegalovirus (CMV) is so ubiquitous that subclinical postnatal infection is very possible, especially in institutions for the retarded. 2. The virus will continue to be shed in saliva and urine for some time, posing a potential hazard to health care workers caring for these children. Another reservoir of the virus is the immunologically compromised patient, such as an organ transplant recipient who often receives much attention from health care providers, many of whom may be expectant mothers. The first steps in diagnosis are the demonstration of anti-CMV antibodies in significant amounts and the demonstration of characteristic cytomegalic inclusion cells in stained urinary sediment, or culture of the CMV from urine or throat swabs. Available data do not indicate any definite risk of recurrence among children born of a mother who previously delivered a child with CID (26). There is currently no place for amniocentesis in evaluating such pregnancies. The meaning of either finding or not finding CMV in the amniocentesis specimen would not be clear. The same is true for all the intrauterine viral disorders.

Herpes simplex virus, type II (HSV-II) infections in the newborn may be acquired in utero, from passage through the birth canal, or from an infected person in the delivery room or nursery.

Intrauterine infections may be associated with prematurity, failure to thrive, developmental delay, hypotonia, convulsions, cutaneous lesions, jaundice, hepatosplenomegaly, anemia, and thrombocytopenia (27). Diagnosis is pursued through maternal and infant serologic testing and culture of the virus. As with CMV, the sooner a diagnosis is made, the better, so that an incidental postnatal infection with this common organism will not confuse the issue. It is also important to distinguish the type-II from the type-I variety for epidemiologic purposes.

Varicella and *mumps* viruses are suspected but not known to cause relatively specific congenital syndromes (28). Other viruses, including those associated with recurrent epidemics of respiratory and gastrointestinal infections, have also been implicated in the etiology of congenital malformations, again without substantiation. The possibility that some cases of *nonprogressive arthrogryposis congenita* (and the hypotonic variant, anarthrogryposis) may be due to maternal–fetal viral infections (29,30) needs to be investigated. The Coxsackie viruses especially are to be considered (31,32). The features of the syndrome include a history of a nonspecific viral illness in mid-pregnancy, the diminution or cessation of fetal movements, and the presence of multiple joint contracture or severe muscular hypotonia at birth, with a subsequent course characterized by progressive improvement if vigorous physical therapy is given. It has been suggested that neurotropic and/or myotropic virus infection of the fetus leads to decreased muscle activity with consequent development of contractures or hypotonia.

As a group, infectious teratogens are not associated with anomalies of the axial or limb skeleton, except for cranial abnormalities (e.g., microcephaly) reflecting aberrant brain growth. Thus, the presence of limb malformations points away from these diagnoses or represents an incidental finding. Intrauterine growth retardation, poor motor development, microcephaly, ocular abnormalities, intracranial calcifications, jaundice, hepatosplenomegaly, anemia, and thrombocytopenia point toward intrauterine infectious teratogenesis. The specific agent should be identified as soon as possible, before postnatal infections can confuse the picture, to afford counseling for recurrence risks and prognosis, and to minimize the chances of infecting other persons.

Chemical and Pharmacologic Teratogenesis

Chemical teratogens are ingested or inhaled inadvertently as pollutants in air, water, or food. Although many compounds are under suspicion in this regard (33), organic mercurials are probably the best documented chemical teratogens (34). The congenital syndrome they produce includes neurologic deficits and blindness (34). Diagnosis depends on a history of exposure to contaminated food (fish) or water and the demonstration of increased levels of tissue mercury.

Pharmacologic teratogenesis is of increasing concern to all clinicians. The number of drugs that can be specified as definite teratogens, or even strongly suspected ones, is relatively small (Table 7-4). But the syndromes they produce clarify several general principles regarding the use of drugs in pregnant women and women who are capable of childbearing. The overriding principle is that *no drug should be administered without a specific intended benefit that cannot be derived in any other way at less risk*. There is no basis for teratogenesis mediated through a father's exposure to a drug (discounting mutagenesis and clastogenesis as distinct mechanisms). Nor is there a basis for a drug *directly* causing congenital abnormalities if pregnancy begins after use of the drug has stopped. The only possible exception is chromosome change of certain types (e.g., triploidy) if conception occurs within 2-3 menstrual cycles after termination of the use of birth control pills (35). Even then, however, the drug would not directly affect the fetus. We will not deal here with the techniques and limitations of teratology investigations except to point out that the amount of the drug and the time of administration are important variables, as is the species specificity of the response. It is tenuous at best to project results from animals to humans.

The "definite hazards" listed in Table 7-4 are clearly established as teratogens. The number is surprisingly small. The most well known is *thalidomide,* a drug no longer available for clinical use. When the drug was taken early in pregnancy, the full range of malformations included phocomelia of all four extremities, eye abnormalities (e.g., colobomata), cutaneous hemangiomas, and congenital heart disease. When taken later in pregnancy, a triphalangeal thumb was sometimes the only abnormality. Between

TABLE 7-4

Drugs Definitely Known or Strongly Suspected of Having Human Teratogenic Affect

A. Definite hazards
   1. Thalidomide
   2. Amethopterin
   3. Progestins (masculinizing effects)
B. Suspected hazards
   1. Anticonvulsants, especially the hydantoins
   2. Warfarin anticoagulants
   3. Ethanol
   4. Progestins
   5. Lithium
   6. Amphetamines
   7. Antitumor agents in general

these extremes there were many combinations of anomalies. The outstanding finding from the compiled data was that when the drug was taken in the critical time period, the likelihood of an aberration of some type was virtually 100%. Another lesson was the consistency of the malformations (taking into account the time of exposure): a distinct syndrome, albeit with several variations, developed.

Amethopterin and *aminopterin* are chemically related drugs which inhibit folic acid metabolism and are useful in treating certain malignancies (e.g., choriocarcinoma, childhood leukemia), psoriasis, and a few other disorders. They are also "street drugs" frequently used as illicit abortifacients. If they do not cause abortion, the fetus suffers severe intrauterine growth retardation, cerebral and cranial malformations, distinctive facies, and diffuse skeletal and visceral anomalies (36).

Progestins, depending on the dosage and timing, can virilize a female fetus. Thus, a risk in treating a threatened miscarriage over prolonged periods with progestins is birth of a female infant with genitalia that are more male than female, if not completely male.

The second group of drugs listed in Table 7-4 are suspected of being teratogenic because some exposed human fetuses manifest a consistent syndrome at birth. The apparent inconstancy could reflect a range of severity of the drug effects or subpopulations of women or fetuses having different susceptibilities. The least that can be said for each of them is that *no pregnant woman taking them can be assured that she is free of risk.*

Hydantoin anticonvulsants are strongly implicated as the cause of a syndrome of the same name which occurs in a proportion of children born to epileptic mothers (37). Whether the underlying disorder itself is a factor remains unclear (38). The "hydantoin syndrome" is characterized by a distinctive facies with telecanthus, epicanthic folds, and cleft palate; by terminal digital hypoplasia (fingers and toes) with associated dermal arches and underdeveloped nails; and by intrauterine and postnatal growth retardation, and delayed psychomotor development (Fig. 7-3). Congenital heart disease is also said to be a component. The pathogenesis is not understood and thus there is no recognized way of preventing the syndrome if a woman must have hydantoin or a similar agent during pregnancy. Every pregnant woman with epilepsy controlled by a hydantoin derivative has a risk perhaps as high as 20% for bearing a child with features of the syndrome. Not only must this risk be weighed against seizure control, but the mother must be fully informed that the risk is there. Perhaps at present this is the main issue: that the woman needs to be informed about the teratogenic potential of the situation so that she and her husband can decide on pregnancy or its continuation. It is their choice, not that of the prescribing physician. If a child

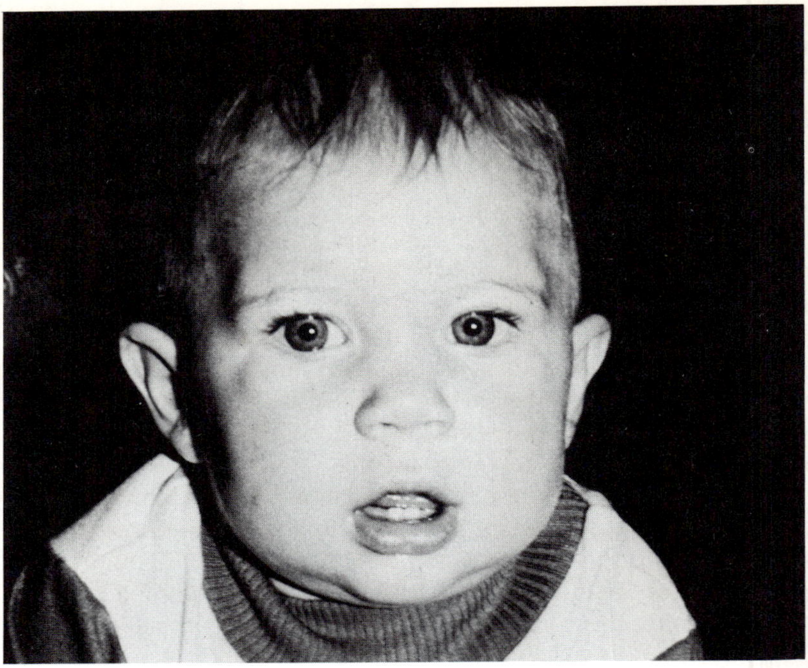

(a)

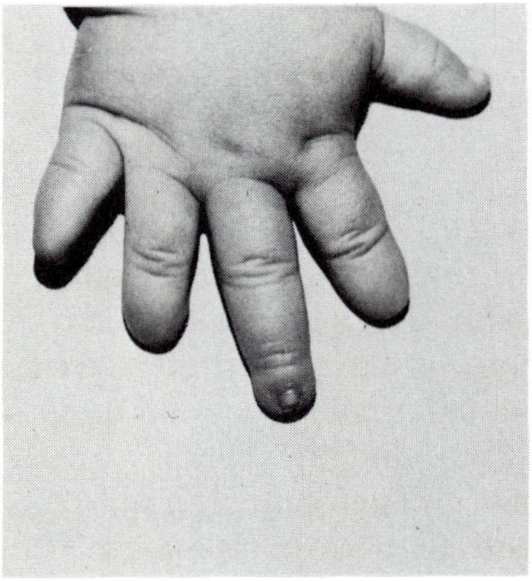

(b)

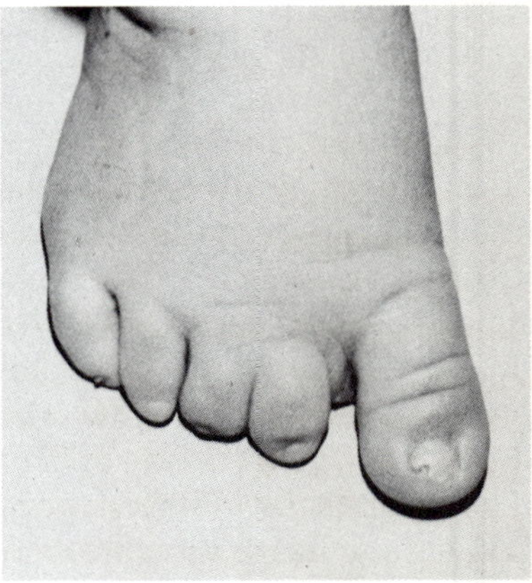

(c)

Figure 7–3 Facies and digital anomalies associated with the fetal hydantoin syndrome. (a) Note the full face, wide-spaced distance between the inner angles of the eyes (telecanthus), and prominent lower lip. (b) There are no nails on fingers II, IV, and V and the one on III is small. All distal phalanges are hypoplastic. (c) As in the hands, all distal phalanges, especially for digits II–V, are hypoplastic.

with the hydantoin syndrome has already been born to a couple, then the risk for future siblings to the child would be even clearer. Families with multiply affected siblings are known to the author.

The term "trimethadione syndrome" which is applied to affected children born of mothers taking this drug for seizure control during pregnancy (39), implies that there may be a separate syndrome for each anticonvulsant agent. This, I believe, remains to be demonstrated.

The anticoagulant drug *warfarin* has recently been associated with characteristic congenital anomalies in children born to mothers who took it during early and middle pregnancy (40,41). Among the features are intrauterine growth retardation, relatively specific types of skeletal abnormalities, including stippled epiphyses and prominent occiput, a distinctive facies, underdeveloped nasal bones, psychomotor retardation, and optic atrophy. The retrospective assignment of etiology can be helpful to parents of an affected child, especially since the drug is ordinarily taken for relatively short periods of time and recurrence in the absence of the drug is not expected. As with other drugs we have discussed, however, it is imperative that every pregnant woman be advised of the teratogenic risk. A pregnancy test before and several weeks after initiating therapy, plus contraception during the period of administration, are reasonable precautions.

Ethyl alcohol, in the context of chronic alcoholism unabated during pregnancy, is definitely a teratogenic hazard. It is included here as a "suspected" human teratogen because it is not yet clear whether alcohol itself is the offending agent or whether the pathophysiologic state of chronic, sustained alcoholism represents the primary insult. The *fetal alcohol syndrome* is characterized by intrauterine and postnatal growth retardation, delayed psychomotor development or frank mental retardation, microcephaly, distinctive facies with epicanthal folds, telecanthus, small palpebral fissures, small mouth, micrognathia, palate abnormalities, dermatoglyphic abnormalities including simian creases, minor skeletal anomalies (e.g., clinodactyly), and congenital heart disease (42,43). The neonatal period is characterized by jitteriness and/or convulsions, which abate with time; it is not clear that these are literal "withdrawal" symptoms. Recognition of the disorder is important in clarifying recurrence risks for other family members. Mutliple affected siblings, including twins, are known (44). Alcoholic women who become pregnant should be informed of the potential teratogenic consequences. It is not known whether abstention upon diagnosis of pregnancy removes or substantially decreases the risk: The nature of the malformations, including those noted at autopsy, suggest that the adverse effects become manifest very early in gestation. An alcoholic woman, informed of these possibilities, may choose to terminate the pregnancy.

It is important to emphasize that we are talking about the disease state of chronic alcoholism. There is no evidence at this time to implicate light social consumption of alcoholic beverages or even isolated episodes of intoxication.

Progestational agents in various forms have been implicated as possible teratogens (apart from their role in producing the virilization effects noted above), but the matter is controversial. It is of special importance, however, in view of the extraordinarily wide use of these agents in birth control pills, preparations to diagnose pregnancy, and preparations to treat certain obstetrical conditions such as threatened abortion. The mere fact that a controversy exists warrants reassessment of the use of these preparations in certain circumstances. Is there still a need for a provocative pregnancy test, especially if a woman has already borne a malformed child (cause unknown) and better diagnostic methods are available? Is it reasonable to treat a "threatened abortion" with progestins if it is *possible* that the treatment will lead to problems *and* its efficacy is subject to debate (45,46)? Is it reasonable to continue progestin-containing birth control pills after a "missed period," recognizing that the phenomenon might represent either a pharmacologic effect of the medication *or* a treatment failure (pregnancy)?

Various types of congenital abnormalities have been ascribed to progestins, ranging from isolated limb-reduction defects (47) to more comprehensive syndromes. For example, the VACTERL syndrome (48,49) involves *v*ertebral, *a*nal, *c*ardiac, *t*racheo-*e*sophageal, *r*enal, and *l*imb anomalies when it is full-blown, or any combination of four or more of them in partial forms of the syndrome. The vertebral anomalies include incomplete closure of the neural arch and segmentation deformities (e.g., extra, missing, or wedge vertebrae). The anal anomaly can be lack of perforation and/or displacement. A wide variety of cardiac defects have been noted. Tracheo-esophageal fistula is the most common form of involvement of those organs. Renal anomalies vary from underdevelopment to fusion or duplication. The limb malformations are usually reduction defects, ranging from hypoplastic phalanges to amelia.

This combination of anomalies is quite nonspecific and thus, it has a number of causes. It may even be part of a broader multiple-anomaly syndrome (including various chromosome disorders). One possible cause for this recurring pattern of anomalies in widely separated organ systems is a particular drug class. It was the recurring pattern of malformations for example, that led to the recognition of the teratogenic liability of thalidomide. Although the spectrum of anomalies is somewhat broader in the VACTERL syndrome, and although the evidence linking progestins to this syndrome is only tentative, there are enough retrospective and prospective data to warrant some special precautions as well as additional study (50).

In general, the diagnosis of the VACTERL syndrome means that recurrence in future pregnancies is not to be expected, although any drug administration in subsequent pregnancies should be avoided.

Part of the problem in assigning cause may reflect the widespread and varied use of progestins. That is, only a subpopulation of all those exposed may have an increased risk. Since the VACTERL syndrome can occur in the absence of any identifiable environmental agent, there may be major intrinsic determinants of at least some cases. External factors, such as progestins, might just represent the final insult in such predisposed persons. The purpose here is not to make a case for progestins as a cause of the VACTERL syndrome, but rather to point out that merely because the percentage of affected individuals among all those exposed to an agent is low, further consideration of the agent as a cause in some cases is not ruled out. In the case of progestins, available data suggest that for the time being the prudent answers to the questions posed on p. 146 would be *no*. They further suggest that a woman who has delivered a child with the VACTERL syndrome and was exposed to a progestin compound in early pregnancy should not be similarly exposed in future pregnancies.

Lithium chloride has been widely used to treat manic-depressive psychosis. In several cases of use during pregnancy, there were major malformations in the offspring, especially cardiac lesions including the Ebstein anomaly, which is otherwise very rare (51). Thus, it would be reasonable to avoid the use of lithium during pregnancy, or at least to advise women of childbearing age of its possible teratogenic effect.

Amphetamines, taken illicitly or licitly, have been claimed to exert a teratogenic effect on cardiac development (52). Although this claim has been disputed (53), it would suggest that amphetamine prescription for obstetrical weight control is not prudent. This possible hazard is also worth mentioning because of all the illicit drugs, including LSD and marijuana, amphetamines are the only ones that have shown a possible pattern of adverse effects.

Antineoplastic agents of all types, with the exception of corticosteroids, should be avoided in pregnancy, and it would be ill-advised to become pregnant during a course of treatment with these drugs (54). This conclusion is based on case reports and the mechanisms of drug action: deleterious effects would be expected from agents acting preferentially on rapidly dividing cells, which, of course, characterize a fetus. The occasional report of an apparently normal newborn infant that had developed in the presence of anticancer treatment does not negate the general sanction against the use of these drugs in pregnancy.

Cigarette smoking is known to be associated with lower birthweights, and the effect is apparently proportional to the severity of the habit (55). However, there have been no reports implicating cigarette smoking as a cause of malformations. For the woman whose obstetrical course and outcome are already at risk (i.e., previous premature birth, Rh incompatibility, the presence of diabetes mellitus) the added insult of cigarette smoking is clearly not desirable.

The number of drugs known or strongly suspected to have teratogenic effects is small in comparison with the myriad drugs available to pregnant women with or without prescription. Table 7-5 lists some of the drugs or types of drugs for which a clearcut teratogenic effect has never been established despite their extensive clinical use. They are presented merely as a contrast to those drugs that are clearly a source of concern.

Maternal Environmental Teratogenesis

The physiological status of the mother can obviously affect the development of a fetus. Hormones and metabolites which cross the placental barrier and other less specific factors may adversely influence embryologic development (Table 7-6).

The role of *maternal malnutrition* in fetal development is not entirely clear. Most analyses fail to find definite teratogenic effects of malnutrition, either in terms of mental retardation or anomalies (56). There is, however, an associated decrease in fertility, fecundity, and infant birthweight (57). Whether severe or prolonged *fever* can be teratogenic is unknown, though it remains a source of concern. *Diabetes mellitus* is known to be associated with obstetrical complications, prematurity, and congenital malformations (e.g., sacral agenesis), although the specific mechanisms are unclear (58). It is not known with any certainty whether careful treatment of hyperglycemia and avoidance of diabetic ketoacidosis significantly alter the risk of major malformations. Increased

TABLE 7-5
Drugs Never Established as Having a Teratogenic or Negative Genetic Effect in Humans

Phenothiazines	Antihypertensives
Diazepam, diazepoxide	Diuretics
Meprobamate	Digitalis preparations
LSD	Corticosteroids
Heroin	Aspirin
Marijuana, hashish	Caffeine
Cocaine	Penicillamine

TABLE 7-6
Maternal Influences Considered to Have Teratogenic Effects

1. Maternal endocrine or metabolic imbalance, e.g., virilizing adrenal tumor
2. Diabetes mellitus[1]
3. Maternal PKU (phenylketonuria)
4. Maternal alcoholism (?)[2]

[1]It is not clear that an infant of a diabetic mother is at risk for all types of malformations, though some types, such as sacral agenesis, are more common when the mother is a diabetic (see text).
[2]The query means that it is unknown whether ethanol *per se* is teratogenic, or whether it is the mother's alcoholic state that causes the congenital defects.

serum levels of *endogenous androgenic hormones*, as for example from a virilizing adrenal or ovarian tumor, can have a masculinizing effect on developing female infants. Any trophic or target hormone which crosses the placenta can presumably disrupt fetal homeostasis and lead to developmental aberrations.

A woman with untreated *phenylketonuria* (PKU) poses a substantial teratogenic threat to her fetus (59,60). This possibility is doubly important because, in some instances, there may be no clinical signs of her disorder. An awareness of this maternal teratogenic effect can be implemented retrospectively and prospectively. Retrospectively, it can account for the clinical picture of mental retardation, microcephaly, and intrauterine and postnatal growth retardation. For infants or children with this picture and no known etiology, it is always reasonable to screen the mother for PKU, especially if there are two or more affected children in a family. Prospectively, women found to have clinically inapparent PKU (detected perhaps by screening siblings of a severely retarded child) should be advised of a high risk for having an affected child. Although some investigators would suggest that there is essentially no significant risk if an affected woman's diet is restricted to bring her serum phenylalanine down to 15 mg % or less, the data are probably too limited to verify that this regimen lowers the risk (61). In addition, the untested effect of a low phenylalanine diet on a fetus would then have to be considered, with special concern for inadvertent episodes of very low serum phenylalanine. We do not know if there are other inborn errors of metabolism that can have serious consequences for developing offspring.

Whether the *fetal alcohol syndrome* should be considered a maternal environmental or pharmacologic effect is an unanswered question that urgently needs to be resolved.

The differential diagnosis of congenital disorders involves both genetic and environmental etiologies. It is important to recognize the clinical similarities and recurrence risk differences between genetic and nongenetic disorders. A diligent effort may be necessary to clarify the environmental etiology of a nongenetic disorder, but the recurrence risks thereby defined will be low so long as the environmental agent is avoided.

Cited References

1. Knudson AG, Hethcote HW, Brown BW: Mutation and childhood cancer: a probabilistic model for the incidence of retinoblastoma. *Proc Natl Acad Sci* 72:5116, 1975.

2. Kalter H: Some relations between teratogenesis and mutagenesis. *Mutat Res* 33:29, 1975.

3. Huberman E: Correlation between mutagenicity and carcinogenecity. *Mutat Res* 29:285, 1975.

4. Harris CC: The carcinogenicity of anticancer drugs: A hazard in man. *Cancer* 37:1014, 1976.

5. Goad WB, Robinson A, Puck T: Incidence of aneuploidy in a human population. *Am J Hum Genet* 28:62, 1976.

6. Smith DW: Dysmorphology (teratology). *J. Pediatr* 69:1150, 1966.

7. Barrow MV, Staffek AJ: Teratologic and other embrotoxic effects of beta-aminoproprionitrile in rats. *Teratology* 10:165, 1974.

8. Philippe JY: Fertility and irradiation: A preconceptional investigation in teratology. *Am J Obstet Gynec* 123:714, 1975.

9. Cleaver JE, Bootsma D: Xeroderma pigmentosum: Biochemical and genetic characteristics. *Ann Rev Genet* 9:19, 1975.

10. Blot WJ, Sawada H: Fertility among female survivors of the atomic bombs of Hiroshima and Nagasaki. *Am J Hum Genet* 24:613, 1972.

11. Warkany J: *Congenital Malformations—Notes and Comments.* Chicago, Year Book Publishers, 1971, pp 38–40.

12. Emerit I, Emerit J: Virus-induced chromosomal breakage. *Lancet* 2:494, 1972.

13. Klein G: The Epstein-Barr virus and neoplasia. *New Engl J Med* 293:1353, 1975.

14. Kolata GB: Estrogen drugs: Do they increase the risk of cancer? *Science* 191:838, 1976.

15. Ziel HK, Finkle WD: Increased risk of endometrial carcinoma among users of conjugated estrogens. *New Engl J Med* 293:1167, 1975.

16. Edmondson HD, Henderson B, Benton B: Liver cell adenomas associated with use of oral contraceptives. *New Engl J Med* 294:470, 1976.

17. Herbst AL, Poskanzer DC, Robboy SJ, et al: Prenatal exposure to stilbestrol. *New Engl J Med* 292:334, 1975.

18. Weinstein D, Mauer I, Solomon HM: The effect of caffeine on chromosomes of human lymphocytes. In vivo and in vitro studies. *Mutat Res* 16:391, 1972.

19. Cohen MM, Hirschhorn K, Frosch WA: In vivo and in vitro chromosomal damage induced by LSD-25. *New Engl J Med* 277:1043, 1967.

20. Taber LH, Huber TW: Congenital Syphilis. in Krugman S, Gershan AA (eds): *Infections of the Fetus and the Newborn Infant.* New York, Alan R. Liss, 1975, pp 183–190.

21. Desmonto G, Couvreur J: Congenital toxoplasmosis, *New Engl J Med* 290:1110, 1974.

22. Alford CA, Stagno S, Reynolds DW: Toxoplasmosis: Silent Congenital Infection. in Krugman S, Gershan, AA (eds): *Infections of the Fetus and the Newborn Infant.* New York, Alan R. Liss, 1975, pp 133–158.

23. Cooper LZ: Congenital Rubella in the United States. in Krugman S, Gershan AA (eds): *Infections of the Newborn Infant.* New York, Alan R. Liss, 1975, pp 1–22.

24. Marx JL: Cytomegalovirus: A major cause of birth defects: *Science* 190:1184, 1975.

25. Hanshaw JB, Scheiner AP, Moxley AW, et al: CNS Sequellae of congenital cytomegalovirus infection. in Krugman S, Gershan AA (eds): *Infections of the Fetus and the Newborn Infant.* New York, Alan R. Liss, 1975, pp 47–54.

26. Yeager AS, Martin HP, Stewart JA: Congenital cytomegalovirus infection: Outcome for the subsequent sibling. *Clin Pediatr* (in press).

27. Nahmias AJ, Visitine AM, Reimer CB, et al: Herpes simplex virus infection of the fetus and newborn. in Krugman S, Gershan AA (eds): *Infections of the Fetus and the Newborn Infant.* New York, Alan R. Liss, 1975, pp 63–78.

28. Siegel M: Congenital malformations following chickenpox, measles, mumps and hepatitis. *JAMA* 226:1521, 1973.

29. Opitz J: personal communication.

30. Peters HA, Opitz J, Goto I, et al: The benign proximal spinal progressive muscular atrophies. *Acta Neurol Scandinav* 44:542, 1968.

31. Hall JG: personal communication.

32. Hall JG, Greene G, Shinkoskey S, et al: Arthrogryposis—Clinical and genetic heterogeneity. *Clin Res* 24:171A, 1976.

33. Smithells RW: Environmental teratogens of man. *Br Med Bull* 32:27, 1976.

34. Apt L, Gaffney WL: Ocular teratogens: A commentary. *J Pediatr* 87:844, 1975.

35. Carr DH: Chromosomal abnormalities in clinical medicine. *Progr Med Genet* 6:1, 1969.

36. Milunsky A, Graef JW, Gagnar MF Jr: Methotrexate induced congenital malformations with a review of the literature. *J Pediatr* 72:790, 1968.

37. Hanson JW, Smith DW: The fetal hydantoin syndrome. *J Pediatr* 87:285, 1975.

38. Shapiro S, Slone D, Harty SC, et al: Anticonvulsants and parental epilepsy in the development of birth defects. *Lancet* 1:272, 1976.

39. Zackai EH, Mellman WJ, Hanson JW: The fetal trimethodione syndrome. *J Pediatr* 87:280, 1975.

40. Shaul W, Emery H, Hall J: Chondrodysplasia punctata and maternal warfarin use during pregnancy. *Am J Dis Child* 129:360, 1975.

41. Pauli RM, Madden JD, Kranyler KJ: Warfarin therapy initiated during pregnancy and phenotypic chondrodysplasia puncta. *J Pediatr* 88:506, 1976.

42. Hanson JW, Jones KL, Smith DW: Fetal alcohol syndrome. *JAMA* 235:1458, 1976.

43. Jones KL, Smith DW: The fetal alcohol syndrome. *Teratology* 12:1, 1975.

44. Christoffel KK, Salafsky I: Fetal alcohol syndrome in dizygotic twins. *J Pediatr* 87:963, 1975.

45. Hellman LM, Pritchard JA: *Williams Obstetrics* (14th ed). New York, Appleton-Century-Crofts, 1971, p 507.

46. Shearman RP, Garrett WJ: Double-blind study of the effect of 17-hydroxyprogesterone caproate on abortion rate. *Br Med J* 1:292, 1963.

47. Janerich DT, Piper JM, Glebatis DM: Oral contraceptives and congenital limb reduction defects. *New Engl J Med* 291:697, 1974.

48. Nora JJ, Nora AH: Birth defects and oral contraceptives. *Lancet* 1:941, 1973.

49. Nora JJ, Nora AH, Perenchief AG, et al: Congenital abnormalities and first trimester exposure to progesterone/estrogen. *Lancet* 1:313, 1976.

50. Harlap S, Prywes R, Davies AM: Birth defects and estrogens and progesterones in pregnancy. *Lancet* 1:682, 1975.

51. Nora JJ, Nora AH, Torews WH: Lithium, Ebstein's anomaly and other congenital heart defects. *Lancet* 2:594, 1974.

52. Nora JJ, Vargo TA, Nora AH, et al: Dexamphetamine a possible environmental trigger in cardiovascular malformations. *Lancet* 1:1290, 1970.

53. Briggs GG, Samson JH, Crawford DJ: Lack of abnormalities in a newborn exposed to amphetamines during gestation. *Am J Dis Child* 129:249, 1975.

54. Mennuti MY, Shepard TH, Mellman WJ: Fetal renal malformation following treatment of Hodgkins disease during pregnancy. *Obstet Gynec* 46:194, 1975.

55. Davies DP, Gray OP, Elwood PC, et al: Cigarette smoking in pregnancy: Associations with maternal weight gain and fetal growth. *Lancet* 1:385, 1976.

56. Stein Z, Susser M, Saenger G, et al: *Famine and Human Development.* New York, Oxford University Press, 1975.

57. Lechtig A, Delgado H, Lasky R, et al: Maternal nutrition and fetal growth in developing countries. *Am J Dis Child* 129:553, 1975.

58. Soler NG, Walsh CH, Malins JM: Congenital malformations in infants of diabetic mothers. *Quart J Med* new series 45:303, 1976.

59. Mabry CC, Denniston JC, Coldwell JC: Mental retardation in children of phenylketonuric mothers. *New Engl J Med* 275:1331, 1966.

60. Angeli E, Denman AR, Harris RF, et al: Maternal phenylketonuria: A family with seven mentally retarded siblings. *Develop Med Child Neurol* 16:800, 1974.

61. Scriver CR, Rosenberg LE: *Amnio Acid Metabolism and its Disorders.* Philadelphia, WB Saunders, 1973, p 329.

General References

A. Beck F: Model systems in teratology. *Br Med Bull* 32:53, 1976.

B. Committee 17, Council of the environmental mutagen society: Environmental mutagenic hazards. *Science* 187:503, 1975.

C. Dudgeon JA: Infective causes of human malformations. *Br Med Bull* 32:77, 1976.

D. Favus MJ, Schneider AB, Stachura ME, et al: Thyroid cancer occurring as a late consequence of head and neck irradiation. *New Engl J Med* 294:1019, 1976.

E. Kelly S, Hook EB, Janerich DT, et al (eds): *Birth Defects. Risks and Consequences.* New York, Academic Press, 1976.

F. Krugman S, Gershan AA (eds): *Infections of the Fetus and Newborn Infant.* New York, Alan R. Liss, 1975.

G. Lewis EB: Possible genetic consequences of irradiation of tumors in childhood. *Radiol* 114:147, 1975.

H. Little JB: Environmental hazards—ionizing radiation. *New Engl J Med* 275:929, 1966.

I. Mims C: Comparative aspects of infective malformations. *Br Med Bull* 32:84, 1976.

J. Rugh R, Budd RA: Does x-radiation of the pre-conceptional mammalian ovum lead to sterility and/or congenital anomalies? *Fertil Steril* 26:560, 1975.

K. Shepard TH: *Catalog of Teratogenic Agents.* Baltimore, Johns Hopkins University Press, 1973.

L. Smith DW: The VATER association. *Am J Dis Child* 128:767, 1974.

M. Soler NG, Walsh CH, Malins JM: Congenital malformations in infants of diabetic mothers. *Quart J Med* (new series) 45:303, 1976.

N. Wilson JC: Present status of drugs as teratogens in man. *Teratology* 7:3, 1973.

CHAPTER EIGHT

PRENATAL DIAGNOSIS

Introduction

Prenatal diagnosis emphasizes the cardinal feature of genetic and congenital disorders—predictability. Given certain circumstances, such as a previously affected child or maternal age of 35 years or more, the risk of having an affected child is sufficiently high to warrant diagnostic procedures aimed at the developing fetus. Decisions based on the diagnosis are a separate issue. With a positive prenatal diagnosis, postnatal therapy and special management could be anticipated, or the parents could elect to terminate the pregnancy. With normal test results, the parents can enjoy a pregnancy that otherwise might have been marred by anxiety.

Prenatal diagnosis has become an established part of health care. The time has passed when it was limited by its newness and experimental nature, or secular sanctions or financial considerations. Many of the general diagnostic precedures used postnatally can be applied in utero: direct inspection; x-rays and ultrasound examinations; analysis of excreted substances (amniotic fluid); and tissue sampling including blood, biopsy specimens, and sloughed cells (suspended in the amniotic fluid). The purpose of this chapter is to describe these applications, provide general guidelines for the use of prenatal diagnosis, and clarify the types of disorders that can be diagnosed prenatally at this time. Several comprehensive reviews of the area are available (1-7). Ethical and legal issues will be briefly mentioned here and discussed in more detail in Chapter 15.

General Clinical Considerations

The most general purpose of prenatal diagnosis is to provide laboratory data about a fetus so that decisions can be made regarding the continuation or outcome of the pregnancy. From the point of view of the clinician, most often an obstetrician, the purpose is to determine whether a fetus in a high-risk pregnancy is afflicted with a given disorder. From the couple's point of view the purpose is either to learn what can be expected if the pregnancy is completed or to be able to decide for or against termination of pregnancy if the fetus is affected with a disorder. Clinicians now have the obligation to inform couples with a high-risk pregnancy that prenatal diagnosis is available. This notion of a *right* to knowledge and use of a diagnostic procedure is relatively new in medicine, and it has developed in part out of controversies surrounding the use of prenatal diagnostic techniques. By far the most specific and frequently used of these techniques is amniocentesis.

Because of the medical, emotional, ethical, and legal gravity of the procedure's possible consequences, genetic amniocentesis must be part of a comprehensive effort which ensures:

1. Respect for the concerns of the couple at all levels.
2. Proper communication of expectations, limitations, and risks of the procedure to allow for informed consent.
3. Proper handling and analysis of the specimen.
4. Proper communication of the results of the analysis.
5. Opportunity to implement any of the alternatives entailed in a positive diagnosis, including termination of pregnancy.
6. Continued followup of mother and child through at least the first year of life.
7. A system within the institution to monitor the operation of the amniocentesis program.

That is, genetic amniocentesis is not merely an obstetric manipulation—it is also a genetic procedure with multiple facets. It is not a screening device. One does not just collect the sample to see whether the results are positive or negative and expect to proceed from there. By collecting the sample we change the couple's attitude toward the pregnancy, and perhaps toward the child-to-be, and efforts must be made to deal with that change (8,9,10).

Specifically, the couple must be part of a system to deal with the results—positive, negative, or equivocal—even before those results are known. If the results are positive: 1. Is this reasonably certain? And what are the checks and controls? 2. Is it certain that all the consequences are clear to the couple? 3. Do they know that the procedure may also be indicated in subsequent pregnancies?

If the results are negative: 1. Is this reasonably certain? Again, what are the checks and controls? 2. Is it certain that something else which represents a substantial risk (i.e., transmission of another genetic disorder) has not been overlooked? 3. Is it certain that the couple understands the limitations of negative results?

For these reasons at the very least, it is essential that amniocentesis be part of a structured program coordinated by an established clinical genetics unit.

Prenatal Diagnosis Techniques

One or more of these techniques may be used, depending on the suspected disorder.

X-ray visualization of the fetus utilizing radio-opaque contrast medium injected into the amniotic fluid may facilitate the diagnosis of certain soft-tissue deformities, such as the protuberance of a myelomeningocele (11). Even without contrast media, anencephaly, hydrocephalus, limb reduction defects, and intrinsic bony defects may be detected (Table 8-1) (12,13). Figure 8-1 demonstrates the intrauterine diagnosis of metatropic dwarfism by radiography. For bony defects the x-rays may be taken beginning at 20 weeks' gestation, using oblique views of the pregnant abdomen. In experienced hands the exposure would be minimal.

Ultrasound examination of the uterine contents is a useful diagnostic procedure in its own right as well as an adjunct to amniocentesis (14,15). As an adjunct, it is used to estimate gestational age, localize the placenta (so that it may be avoided if possible by the amniocentesis needle) (Fig. 8-2), determine the presence of twins (1 in 80 births), and help diagnose neural tube defects. As a primary diagnostic technique ultrasound is mostly applicable to disorders associated with failed mineralization or formation of the skull or enlargement of the skull (e.g., hydrocephalus) (Table 8-2).

Fetoscopy, or direct visualization of the fetus, is a potentially useful technique (16,17) that is not currently available except on an investigative basis because of the apparently high risk of disrupting the pregnancy (17,18). Through a small puncture wound in the abdomen a solid or fiber-optic system can be introduced into the uterine cavity to visualize fetal parts. This direct visualization offers a capacity for the diagnosis of birth defects that cannot be demonstrated indirectly. In addition to the risks of the procedure, however, there is the problem of deciding what

TABLE 8-1
Disorders Amenable to Prenatal Diagnosis by Intrapartum X-rays

1. Disorders associated with limb reduction defects.
2. Disorders characterized by diffuse, early disturbance of bone growth, e.g., metatropic dwarfism.
3. Neural tube defects with and without use of intraamniotic contrast medium (amniography, fetography).

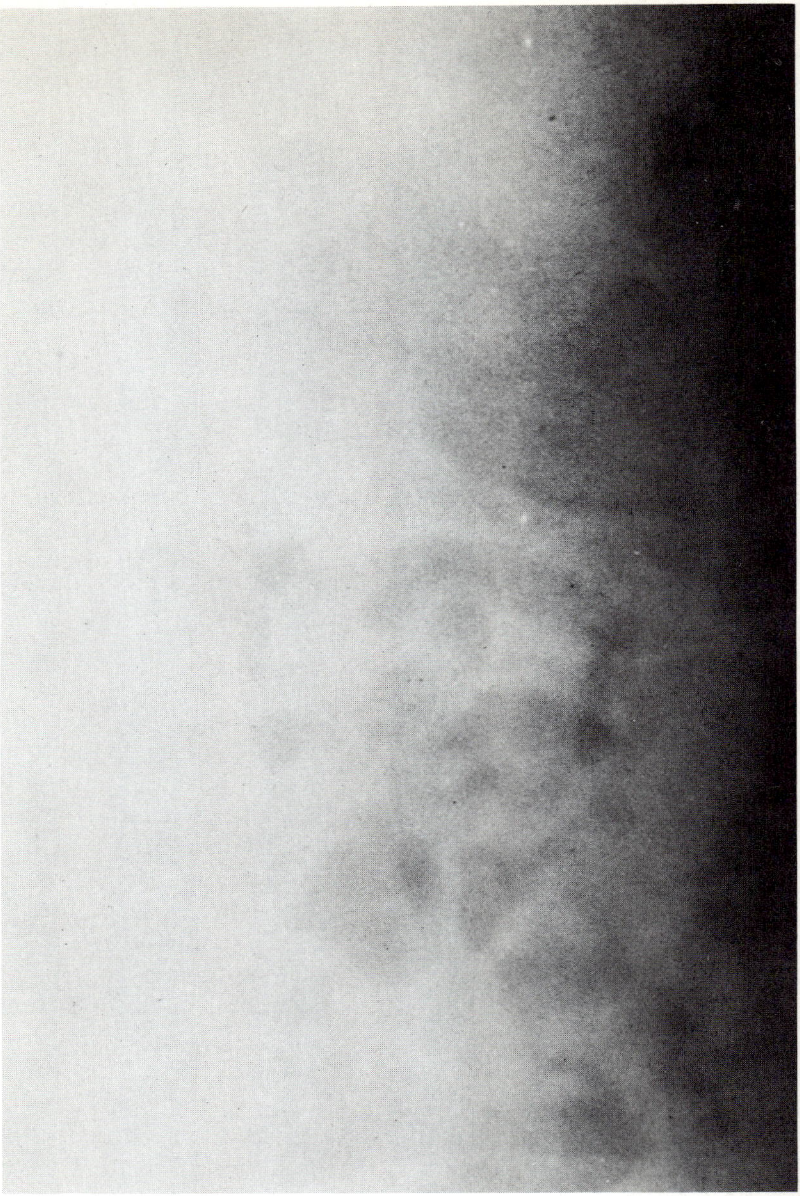

Figure 8–1 Intrauterine diagnosis of metatropic dwarfism by oblique abdominal radiography. The mother's vertebral bodies are on the left. (Photographic contrast has been exagerated for the sake of emphasizing the fetal bones.) The fetal "dumb-bell" shaped femora are visible in the center of the photograph. This appearance indicates metatropic dwarfism. (Courtesy of H. Lubs and D. Harper.)

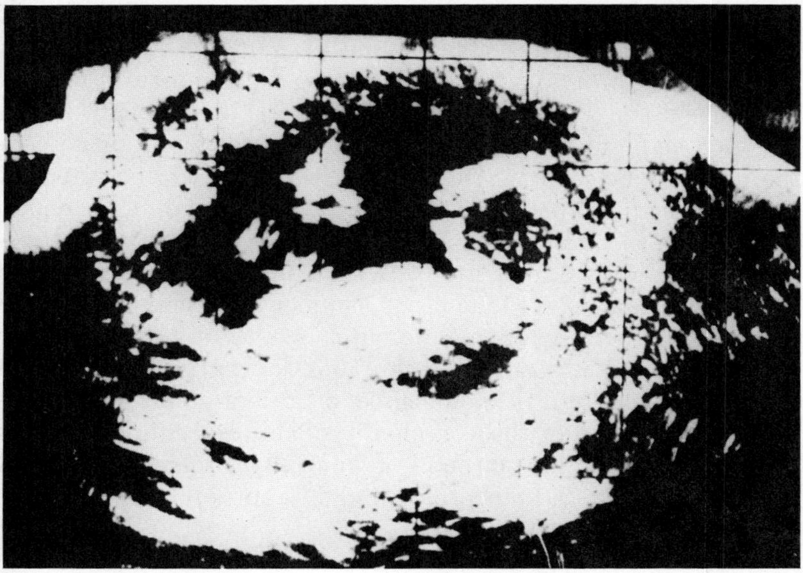

Figure 8–2 Abdominal ultrasound examination at 15 weeks' gestation as part of a prenatal diagnosis evaluation. The fetal head is seen on the right and the placenta on the left. This is a normal scan for a 15-week pregnancy. (Courtesy of K. Gottesfeld.)

would be a sufficiently severe congenital malformation to warrant termination of a pregnancy, who would decide, and by what criteria. For example, would an apparently isolated cleft lip be grounds for ending the pregnancy? If the answer is no, has another factor been introduced when the parents learn of the cleft lip halfway through the pregnancy? Could foreknowledge of a relatively minor defect increase its burden? While many clinical geneticists are enthusiastic about the potential of this technique, we also have reservations about its risks and consequences.

Amniocentesis for prenatal diagnosis is performed at about 16

TABLE 8-2
Genetic Disorders Amenable to Prenatal Diagnosis by Ultrasound

1. Neural tube defects, especially anencephaly
2. Hydrocephalus
3. Hypophosphatasia

Note: Anencephaly would be recognized by absence of cranial echoes, the brain and skull being virtually absent; a large myelomeningocele might be identified as a dorsal cystic mass. Hydrocephalus would manifest as a large cranium. Hypophosphatasia would be suggested by absence of cranial bone echoes (simulating anencephaly) because of deficient skull bone calcification. With continued refinement and experience, diagnosis of renal cystic disease and similar disorders may be routinely diagnosed in utero by ultrasound.

weeks of pregnancy (dated from the first day of the last menstrual period) by withdrawing about 20 ml of amniotic fluid from the amniotic sac through a transabdominal needle puncture using a local anesthetic (Fig. 8-3) (4,19). Biochemical or serologic studies can then be performed on the fluid itself, or the amniotic cells in the fluid can be separated and processed directly or cultured, in either case, for biochemical, immunologic, and/or cytogenetic markers. Most commonly, the fluid is examined directly for alpha-fetoprotein levels and/or the cells are cultured for chromosome analysis. In some cases fetal blood samples are obtained from the placental circulation. As mentioned above, these procedures are combined with ultrasound techniques for guidance (placenta localization) and aid in diagnosis (e.g., absent cranium in anencephaly). Because of the complexity of the procedure's various aspects, coordination through a specially organized clinical genetics program is both professionally appropriate and legally prudent.

The general indications for utilization of this technique are itemized in Table 8-3. *Maternal age* is the most common indication, and its underlying risk is uniquely independent of family history, demonstrable parental mutation (i.e., heterozygosity), or cytogenetic aberration (e.g., balanced reciprocal translocation). Definition of the family history is still necessary when the primary indication is maternal age, since frequently another risk, perhaps even greater, becomes apparent. The age level now used as a routine indication for amniocentesis is 35 years (at time of delivery). Although arbitrary, this cutoff is based on data demonstrating a 1% (1/100) minimal risk for women at or above that age to bear a child with trisomy 21, and an even higher cumulative risk for all types of aneuploidy (20). Moreover, the risk increases with age, so that between 40 and 45 years the risk for trisomy 21 is 1–2% and at 45 years and older, 2.5% or more.

A *previous child with aneuploidy*, regardless of the parents' ages, increases the risk to about 1% (1,7) for recurrence of the same or another form of aneuploidy. If maternal age was not a factor in the first case, then it is not usually clear how the aneuploidy was derived (whether paternal, maternal, or zygotic), and thus amniocentesis is indicated for any subsequent pregnancy involving either of the first case's parents. Sometimes the parental source can be surmised from comparison of the affected child's karyotype with parental karyotypes when special attention is paid to chromosome markers (e.g., prominent satellites). At other times specific parental contribution may be suggested by a positive history in second-degree or other relatives to the proband. The occasional family with such a positive pedigree, consistent with a heritable predisposition to aneuploidy, has even prompted some

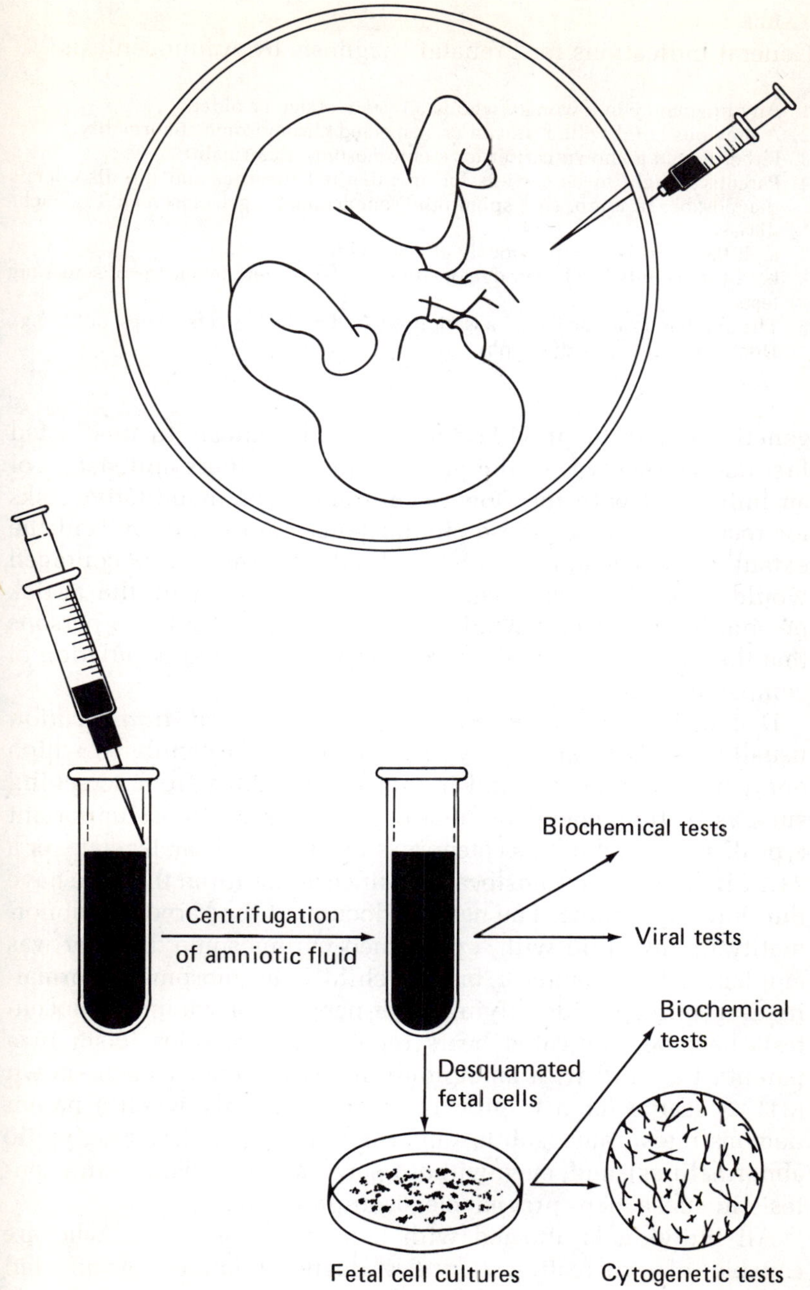

Figure 8–3 Amniocentesis—obtaining and processing the specimen.

TABLE 8-3
General Indications for Prenatal Diagnosis by Amniocentesis

1. Any pregnancy in a woman who is 35 years of age or older.
2. A previous child with a proven or suspected chromosome abnormality.
3. Either parent a known carrier for a chromosome abnormality.
4. Parents thought to be carriers for an inherited disorder and the disorder is diagnosable in utero, e.g., spina bifida/anencephaly, galactosemia, Tay-Sachs disease:
   a. if there has been a previously affected child;
   b. if the parents have been demonstrated to be carriers by means of screening tests.
5. The mother is a carrier of a serious disorder that affects males only, e.g., Duchenne muscular dystrophy.

genetic centers to provide genetic amniocentesis to unaffected first-degree relatives of the proband (e.g., brothers and sisters of an individual with the Down syndrome). The quantitative risks for recurrence under these circumstances are not clear, and the extent to which amniocentesis should be thereby encouraged would depend on the degree of concern shown by the at-risk person. In any case, it would be imprudent to tell these persons that their risk is merely the same as for the general population of comparable age.

Documentation of *mosaicism* or a *balanced translocation* usually results from karyotyping members of a family in which one person is affected, and it sometimes results from screening surveys of the general population. Perhaps the most important type of aberration in this category is the balanced carrier state for a 21;21 Robertsonian translocation, since all children that live have the Down syndrome. The need to document the karyotype abnormality in any child with a presumed chromosome disorder was emphasized in Chapter 3, but the child's chromosome abnormality as a source for identifying other persons for whom amniocentesis becomes indicated bears repetition here. Mosaicism in a parent of a child with aneuploidy has occasionally been shown (21). A search for it is probably warranted only when a parent demonstrates some subtle abnormalities (e.g., dermatoglyphic abnormalities), and, even without parental mosaicism, amniocentesis in subsequent pregnancies is appropriate.

All individuals affected with a *neural tube defect* who are capable of procreating and first-degree relatives of affected patients are considered to be candidates for amniocentesis. Levels of amniotic fluid alpha-fetoprotein at or beyond 3 standard deviations from the normal mean at 14–16 or so weeks of gestation may indicate a neural tube defect (5,22). A normal level may be found in about 10% of affected cases, but then the defect is likely to be smaller and thus clinically less severe. Complementary ultrasound examination is warranted in all cases and fetal x-rays with contrast medium are appropriate in equivocal cases. Since amniotic fluid alpha-fetoprotein levels decrease later in pregnancy, even when a neural tube defect is present, proper timing of the amniocentesis is important. The value of maternal serum alpha-fetoprotein levels in screening for fetal neural tube defects is not yet established (23), but it may eventually become a routine part of prenatal care. Positive serum screening results would need to be corroborated by amniocentesis and/or ultrasound data. Amniotic fluid alpha-fetoprotein levels are checked routinely in many centers, even when the indication for amniocentesis had nothing to do with neural tube defects (24).

In addition to polygenic neural tube defects (Chapter 5), other genetic disorders associated with neural tube defects can be investigated in utero via amniotic fluid alpha-fetoprotein levels (e.g., the Meckel syndrome). Omphaloceles and other disorders of diverse types, such as the Finnish type congenital hydronephrosis, can also be diagnosed in this manner (25).

A moderate number of specific genetic *biochemical disorders* can be diagnosed in utero (26). They are usually autosomal recessive or X-linked recessive traits that can be assessed directly in terms of the primary defect (e.g., enzyme deficiency, hemoglobin abnormality) although for some a secondary aberration is measured (e.g., pyrophosphate levels for osteogenesis imperfecta). For a recent list of biochemical disorders diagnosable in utero the reader is referred elsewhere (26). Because each of these disorders is rare and because the list keeps growing, it is better to emphasize a general principle: if a biochemical disorder is known or suspected, referral to a genetics center is imperative to substantiate the clinical diagnosis, specify the precise biochemical defect (heterogeneity could otherwise be troublesome), indicate the availability of amniocentesis, and advise as to possible therapy. The purposes of intrauterine biochemical diagnosis are: 1. to allow for appropriate treatment as soon after delivery as possible; 2. to allow for termination of pregnancy if so desired by the parents; or 3. to allow for a less emotionally charged pregnancy if no disease is diagnosed.

The complexity of the biochemical assays usually means that each will be available in only one or two laboratories in an area. The initial contact and liaison with the laboratory is the responsibility of the genetics program sponsoring the amniocentesis; the need for special coordination underscores the need for early, comprehensive, in-depth involvement of the geneticist with the patient. The technical complexities of the assay procedures may also introduce unique and troublesome types of false-negative or false-positive diagnoses. These include different enzyme activities for different stages of pregnancy and different results with different alleles for the locus in question (27).

A small number of biochemical disorders can be diagnosed in utero by taking a sample of fetal tissue, usually blood. The blood can be obtained from a fetal vein via fetoscopy, or from mixed maternal–fetal blood in the placenta. Sampling fetal blood by either means is the only way at present to diagnose fetal hemoglobinopathies (e.g., sickle-cell anemia, thalassemia) (28). This approach is still investigative.

A specific type of indirect in utero diagnosis makes use of *genetic linkage:* a disorder that cannot itself be detected is diagnosed by measuring for a trait that is linked to it (see Chapter 9). Thus, if the mutant allele for the disorder in question is on the same chromosome as a measurable allele at a second locus, the presence of the mutant allele at the first locus can sometimes be surmised by determining the presence of the gene product of the second locus. This linkage approach will probably have its greatest use with dominant disorders, since they ordinarily cannot be detected by any other method in utero (with perhaps one or two exceptions) and they have a 50% transmissability risk from an affected parent. The approach has already been employed successfully for at least one autosomal dominant trait, myotonic dystrophy (29,30). There are a number of technical limitations intrinsic to this approach, so that it cannot be used in every family where a given disorder arises, especially if it occurs in the form of a new mutation.

A special type of in utero diagnosis involves cytogenetic *sex determination* to identify male fetuses that would be at-risk for developing disabling or lethal X-linked recessive disorders, such as Duchenne muscular dystrophy, hemophilia A, and the Renpenning mental retardation syndrome. If a woman is a known carrier, each son would have a 50% risk for the disorder. If the mother's carrier status is not certain, the risk for a son is proportionately less. Some families have made use of this approach even when the risk for an affected male is only 1–2%, which emphasizes the importance of a disorder's burden in making such decisions. That is, it is not merely a matter of balancing an amniocentesis-associated miscarriage risk of no more than 1% against a 1–2% disease risk; the severity and duration of the disorder must also be considered. This factor cannot be expressed in numbers and it is a matter of personal judgement for the family involved. For instance, a woman who grew up with a brother afflicted with Duchenne muscular dystrophy may want to do anything she can to avoid having such a son—even if it means having no sons.

Currently, the most common indications for amniocentesis, in decreasing order, are maternal age, a previous child with a chromosome disorder, and a close relative with a neural tube defect. In the long run, neural tube defects will generate the largest number of potential referrals, since for each case of the disease so many others are ultimately identified as being at risk. Among the disorders for which amniocentesis is not indicated are those where the defect is not reflected in the amniotic fluid or cells. These include many biochemical disorders, all heritable disorders without a biochemical or chromosomal cause or marker, and virtually all the congenital malformations except those associated with elevated alpha-fetoprotein levels. The limits of the procedure are probably overshadowed, however, by the limits on making the procedure uniformly available (26).

Amniocentesis Risks

The theoretical risks of amniocentesis include damage to the fetus, unexpected miscarriage, and other obstetrical complications such as aminotic fluid leakage and infection. The cumulative experience, including that from a national collaborative study (31,32), indicates a minimal risk, if any, to mother or fetus when the procedure is part of a large, active clinical genetics program. There is only circumstantial evidence to suggest actual fetal trauma during amniocentesis for prenatal diagnosis at 14–16 weeks gestation (33). The risk of unexpected miscarriage following amniocentesis is essentially the same as for women in a control group, the actual difference amounting to one-half a percentage point (32). To ensure that as advocates of this approach to preventive medicine we are not underestimating a real risk, clinical geneticists ordinarily quote "roughly a 1% risk" to couples considering the procedure. Intrauterine deaths and stillbirths are apparently not increased (32,34).

Other types of complications include inconsequential amniotic fluid leak into the vagina, the possibility of infection at the site of abdominal puncture, and the possibility of Rh antibody formation in an Rh-negative mother carrying an Rh-positive fetus. This last problem can apparently be avoided by administering Rh-immune globulin after amniocentesis, though its use in this regard has only been experimental to date. No risk to mother or fetus has been attributed to the use of ultrasound (35).

Amniocentesis—Technical Considerations

Proper timing of the procedure plus genetic evaluation and counseling are critical to ensure optimal laboratory analysis, minimal patient anxiety while awaiting results, and maximal opportunity to act on positive findings. Referrals should be made as soon as possible after pregnancy is diagnosed in an at-risk woman to allow for the amniotic tap between 14 and 16 weeks gestation. In that time period there is a sufficient amount of amniotic fluid, the cells therein grow optimally, the alpha-fetoprotein levels are significant, and, even with a 3–5 week delay till results are available, there is time to safely terminate the pregnancy if indicated. As previously noted, the geneticist's role in pre-amniocentesis evaluation, (apprisal of expectations, liaison while awaiting the results, support throughout the pregnancy and postdelivery followup) is as much a part of any genetic amniocentesis as the tap itself. The obstetrician must be experienced in performing the tap: the low risk figures are based on the procedure's being performed by clinicians with demonstrated ability and experience. Ultrasound is also a requisite for the reasons noted above. Ultrasound does not appear to have an adverse effect on the viability of the sampled cells or their intrinsic properties, including chromosomes (35). When twins are occasionally diagnosed by ultrasound, both amniotic cavities have to be sampled. This is easily done by sampling one, injecting a dye into it and sampling the other, which is recognized as such by the absence of dye (36). Occasionally, usually for unknown reasons, the cells fail to proliferate and a repeat amniocentesis is necessary; a decision to repeat the tap because of poor cell growth is made within 7–10 days to minimize loss of time before final results are available.

This approach to diagnosis is not yet applicable to fetal infestation with an infectious teratogen. Although viruses could be cultured from the fluid or cells, it is not possible to equate either their presence with fetal disease or their absence with no fetal disease. In several medical centers direct fetal specimens are obtained by combining fetoscopy and amniocentesis, but only in special circumstances as part of a research protocol. Amniocentesis is not currently being used for the purpose of sex selection when no sex-restricted genetic disease is involved.

From the patient's point of view we should appreciate the need for proper informed consent, close communication and followup, and the cost for the whole process. It is imperative that the couple understand the indications, certain technical details, expectations, and limitations of each component of the process. This information must be given with sensitivity to the turmoil that may be experienced by the family. The dollar costs for all components usually are between 300 and $400. What it costs the couple, or their insurance company, depends on the availability of subsidies through service grants or government agencies to the genetic program. Many insurance companies will cover the various costs, the largest of which is for the culture and analysis of the amniotic fluid cells.

The availability of facilities and personnel to perform an abortion is an important consideration, since this procedure may be decided upon when the amniocentesis laboratory data are analyzed and the results presented. Except to note that, in capable hands, the risk involved in an induced abortion during the first half of pregnancy is relatively low (especially in comparison to the burden avoided), the procedure is beyond the scope of this discussion. For more information the reader is referred elsewhere (37,38). The technical aspects of induced abortion may not restrict the availability of genetic amniocentesis as much as the moral or religious convictions of clinicians. Clearly, a clinician's personal convictions should not prevent his patients from having access to the prenatal diagnosis by proper referral. Prenatal diagnosis is not to be equated with abortion; it is merely a means of acquiring data so that a number of decisions can be made, one of which *may* include abortion (39). Moral decisions about abortion must be made by the couple involved, not by the clinicians assisting them. A clinician would be remiss—professionally, ethically, morally, and legally—to deny a couple the opportunity to obtain information about the fetus or to implement a decision about continuing the pregnancy (40).

If an abortion is decided upon as a result of amniocentesis-derived data, the abortus should always be examined to confirm the amniocentesis findings. This is important both for monitoring the accuracy of the diagnostic tests and for assuring the parents.

Cited References

1. Milunsky A, Atkins L: Prenatal diagnosis of genetic disorders. *JAMA* 230:232, 1974.

2. Atkins L, Milunsky A, Shahood JM: Prenatal diagnosis: Detailed chromosomal analysis in 500 cases. *Clin Genet* 6:317, 1974.

3. Laxova R, Lewis BV, Suddaby M: A clinical service for prenatal diagnosis. *Lancet* 2:964, 1975.

4. Schwarz RH: Amniocentesis. *Clin Obstet Gynec* 18:1, 1975.

5. Milunsky A, Atkins L: Prenatal diagnosis of genetic disorders in Milunsky A (ed): *Prevention of Genetic Disease and Mental Retardation*, Philadelphia, WB Saunders, 1975, pp 221–263.

6. MacVicar J: Antenatal detection of fetal abnormality—physical methods. *Br Med Bull* 32:4, 1976.

7. Laurence KM, Gregory P: Prenatal diagnosis of chromosome disorders. *Br Med Bull* 32:9, 1976.

8. Allen HH, Sergovich F, Stuart EM, et al: Infants undergoing antenatal genetic diagnosis: A preliminary report. *Am J Obstet Gynec* 118:310, 1974.

9. Robinson J, Tennes K, Robinson A: Amniocentesis: Its impact on mothers and infants. *Clin Genet* 8:97, 1975.

10. Duncan PE, Finley SC, Finley WH: Impact of prenatal genetic diagnosis on participating couples. *Clin Res* 24:64A, 1976.

11. Suzomorik, Yagami Y: Diagnosis of human fetal abnormalities by fetography. *Teratology* 12:303, 1975.

12. Lubs HA, Lubs ML: Genetic disorders. in Burrow GN, Ferris TS (ed): *Medical Complications During Pregnancy*. Philadelphia, WB Saunders, 1975, pp 810–895 (See Figure 20-9, and pages 864–866).

13. See ref. 6, pages 5 and 6, and Plate I.

14. See ref. 6, pages 6 and 7, and Plate II.

15. See ref. 4, page 10.

16. Laurence KM, Prosser R, Rocker I, et al: Hirschsprung's disease with congenital heart malformation, broad big toes and ulnar polydactyly in sibs: A case for fetoscopy. *J Med Genet* 12:334, 1975.

17. Patrick JE, Perry TB, Kinch RAH: Fetoscopy and fetal blood sampling: A percutaneous approach. *Am J Obstet Gynec* 119:539, 1974.

18. See ref. 5, page 243.

19. Pernoll ML, Prescott GH, Hecht F, et al: Prenatal diagnosis: practice, pitfalls and progress. *Obstet Gynec* 44:773, 1974.

20. Goad WB, Robinson A, Puck TT: Incidence of aneuploidy in a human population. *Am J Hum Genet* 28:62, 1976.

21. Papp Z, Csecsei K, Skapinyecz J, et al: Paternal normal/trisomy 21 mosaicism as an indication for amniocentesis. *Clin Genet* 6:192, 1974.

22. Milunsky A: Prenatal diagnosis of neural tube defects. *Obstet Gynec* 48:1 and 6, 1976.

23. Leighton PC, Gordon YB, Kitau MJ, et al: Levels of alpha-fetoprotein in maternal blood as a screening test for fetal neural-tube defects. *Lancet* 2:1012, 1975.

24. Milunsky A, Alpert E: Routine testing for alpha-fetoprotein in amniotic fluid. *Lancet* 1:1015, 1976.

25. Brock DJH: Mechanisms by which amniotic fluid alpha-fetoprotein may be increased in fetal abnormalities. *Lancet* 2:345, 1976.

26. Milunsky A: Current concepts in genetics. Prenatal diagnosis of genetic disorders. *New Engl J Med* 295:377, 1976.

27. Wenger DA, Riccardi VM: Possible misdiagnosis of Krabbe disease. *J Pediatr* 88:76, 1976.

28. Kan YW, Golbus MS, Trecartin R: Prenatal diagnosis of sickle-cell anemia. *New Engl J Med* 294:1039, 1976.

29. Omenn GS, Schrott HG: Prenatal prediction of autosomal dominant diseases by linkage studies. Myotonic dystrophy. *Acta Genet Med* 23:221, 1973.

30. Schrott HG, Omenn GS: Myotonic dystrophy: Opportunities for prenatal prediction. *Neurology* 25:789, 1975.

31. Culliton BJ: Amniocentesis: HEW backs test for prenatal diagnosis of disease. *Science* 190:537, 1975.

32. Milunsky A: Risk of amniocentesis for prenatal diagnosis. *New Engl J Med* 293:932, 1975.

33. Broome DL, Wilson MG, Weiss B, et al: Needle puncture of fetus: A complication of second trimester amniocentesis. *Am J Obstet Gynec* 126:247, 1976.

34. Robinson A, Bowes W, Droegemueller W, et al: Intra-uterine diagnosis: Potential complications. *Am J Obstet Gynec* 116:937, 1973.

35. Abdulla U, Campbell S, Dewhurst CJ, et al: Effect of diagnostic ultrasound on maternal and fetal chromosomes. *Lancet* 2:829, 1971.

36. Bang J, Nielsen H, Philip J: Prenatal karyotyping of twins by ultrasonically guided amniocentesis. *Am J Obstet Gynec* 123:695, 1975.

37. Wentz AC, Burnett LS, King TM: Methodology in premature pregnancy terminiation, Part I. *Obstet Gynec Survey* 28:2, 1973.

38. Burnett LS, Wentz AC, King TM: Techniques of pregnancy termination, Part II. *Obstet Gynec Survey* 29:6, 1974.

39. Elias S, Mahoney MJ: Prenatal diagnosis of trisomy 13 with decision not to terminate pregnancy. *Obstet Gynec 47 (Suppl)*:75, 1976.

40. Jacobs v. Theimer: *Texas Supreme Court Journal* 18:222, 1975.

General References

A. Conley R, Milunsky A: The economics of prenatal genetic diagnosis. in Milunsky A (ed): *The Prevention of Genetic Disease and Mental Retardation.* Philadelphia, WB Saunders, 1975, pp 442:455.

B. Hazard S, Carter FA: Preventing the birth of infants with Down's syndrome: A cost-benefits analysis. *Br Med J* 1:753, 1976.

C. Littlefield JW, Milunsky A, Jacoby LB: Prenatal genetic diagnosis: Status and problems. in Hilton B, et al (eds): *Ethical Issues in Human Genetics.* New York, Plenum Press, 1973.

D. Milunsky A: *The Prenatal Diagnosis of Hereditary Disorders.* Springfield, IL, Charles C. Thomas, 1973.

E. Milunsky A: Current concepts in genetics: Prenatal diagnosis of genetic disorders. *New Engl J Med* 295:377, 1976.

CHAPTER NINE

COMPONENTS OF A GENETIC EVALUATION

Introduction

"Genetic evaluation" is a shorthand phrase for "clinical evaluation from a genetic point of view." A genetic evaluation is an extension of general medical principles, with acknowledgment of the family as the focus of health care and of the critical importance of diagnosis based on etiology, as discussed in Chapter 2. It can be initiated by any health care professional, although many diagnostic studies must be arranged by physicians. The purpose of this chapter is to delineate the components of a genetic evaluation, demonstrate their application, and provide guidelines for their use in commonly encountered situations.

The Nidus Concept of Genetic Evaluation

The clinical geneticist has two sets of related tasks. First, the disorder in question must be diagnosed as specifically as possible, thereby defining etiology and perhaps pathogenesis, which in turn provides a basis for ascribing risks and prognosis. Second, through counseling, the features of the disorder and the recurrence risks must be communicated to the family (Chapter 10). Before counseling can be meaningful, evaluation must be completed. This simple fact is often overlooked. Since a genetic evaluation often requires input from experts in several laboratory and clinical disciplines, the effort may become complex and prolonged. The clinical geneticist responsible for the case then comes to serve as a focal point for collection of the various data: a *nidus* for the accumulation of data that are then put to use on behalf of the family in consultation. The data are derived from detailed history taking, detailed examination of the proband and family members, laboratory analyses, consultation with other experts, and review of the pertinent literature. The methods of obtaining relevant information from each of these sources will be reviewed in the following pages to give an overview of how it is done, in general, and to stress the importance of gathering needed information before it becomes unavailable (e.g., before the death of a severely affected child).

Family History

Among the types of histories to be obtained (Table 9-1), the family history is especially important in a genetic evaluation. A family history reveals the pattern of a disorder's occurrence in a family and identifies at-risk individuals who should be further evaluated and/or informed of the risks.

A positive family history may indicate a heritable disorder and perhaps even suggest a diagnosis. And it may provide the basis for defining recurrence risks, especially for disorders without a specific diagnosis. For example, multigeneration involvement with male-to-male transmission indicates an autosomal dominant pattern, regardless of the disorder's name. The absence of other affected family members (which some would call a "negative family history") may be compatible with all patterns of inheritance: For dominants or X-linked recessives, the proband may reflect a new or recent mutation; for autosomal recessives and polygenics, there often are no previously affected relatives. Table 9-2 lists five reasons to account for the apparent absence of other affected relatives. This list indicates other things that can be sought in the family history, even if no other members are apparently affected: adoption or nonpaternity; the age of the proband's father for new dominant mutations or the age of the proband's maternal grandfather for new X-linked recessive mutations (1); consanguinity and/or common ethnic backgrounds for autosomal recessive and polygenic traits. That is, the family history may be informative even with only one affected person. A family history is truly negative only when it is noninformative in addition to there being no other affected persons. Such a negative history may become positive when it is taken the second or third time, as repeated efforts lead to recall of forgotten facts. Further, a family history—negative or positive for the proband's condition—may

TABLE 9-1
Types of Historical Data

I. Family history	III. Postnatal
II. Maternal–fetal histories	A. Neonatal
A. Conception	B. Developmental
B. Gestation	C. Personal
C. Delivery	D. Social

TABLE 9-2
Reasons That a Family History May Be Apparently or Actually Negative for Affected Family Members Other Than the Proband

1. Inadequate history, e.g., adoption, poor techniques
2. Nonpaternity
3. New Mutation
4. Variable expression or masking of the trait
5. Autosomal recessive or polygenic traits, especially with small-sized families

Note: This table emphasizes that a "negative" family history does *not* rule out a genetic basis for a proband's disorders.

indicate persons afflicted with other disorders that provide a context for, or even supersede in importance, the referral disorder.

When a family history is positive for the disorder in question and enough individuals are affected, not only is the heritable pattern obvious, but the nature of the disorder for this particular family can be appreciated. For example, the age of onset of a disorder may vary widely in general but have a narrower range within any given family. Also, the manner of presentation may be distinctive, and, in any case, is a guide to determining whether other family members are affected. The course of a disease in a given family will also suggest the medical, emotional, and social burden for both affected and unaffected family members. In some families with Huntington chorea, for example, the onset is in the teens or early twenties and the course is relatively rapid and characterized by rigidity; freedom from symptoms at age 35 probably portends absence of the mutant gene. However, in most affected families, the onset is in the thirties or forties, the course is somewhat slower (with more choreiform movements than rigidity), and a family member would have to be symptomless to beyond age 55 to take comfort (though not certainty) in the conclusion that he would not be affected. Another reason for delineating the details of each family member's manner of affliction is the tremendous variability in expression that is typical of many dominant disorders. An affected person may show only one or two features, a sibling one or two others, and a parent still others; only the composite for all family members indicates the characteristic syndrome. This phenomenon is well known for the Waardenburg syndrome: one person may show the white forelock, another deafness, another hypertelorism, etc. Families are often seen for genetic counseling whose compound features determined through the family history establish the diagnosis.

Table 9-3 summarizes the information that can be gleaned from a carefully taken family history. It is most usefully presented in

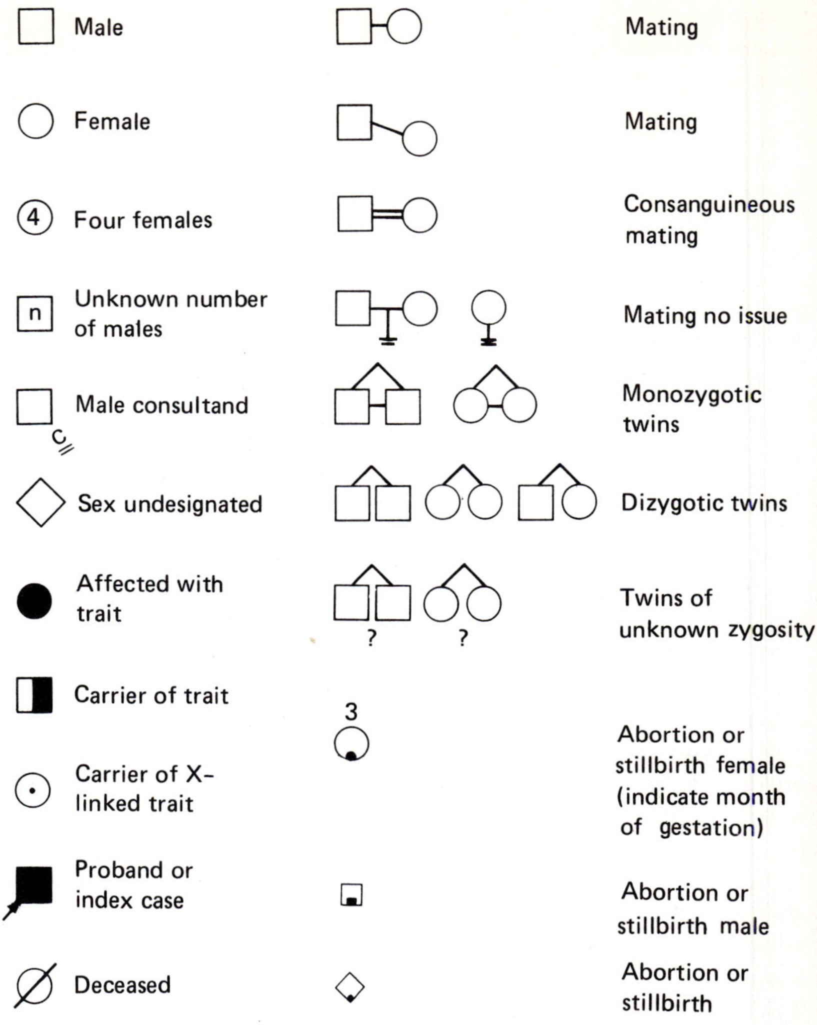

Figure 9–1 Symbols utilized in constructing pedigrees. Note that the consultand is depicted distinctly from the proband (arrow), although sometimes the consultand may be the proband. Also, the carrier for autosomal recessive traits (half-shaded square or circle) is depicted differently from carrier for X-linked trait (circle with central dot). The sizes of abortion and stillbirth symbols are diminutive.

the form of a pedigree or family tree. Figure 9-1 illustrates the symbols used in constructing a pedigree, and Fig. 9-2 an idealized pedigree. Table 9-4 itemizes the rules and guidelines for actually constructing pedigrees. There are two types of pedigrees—one for screening and one for in-depth analysis. The *screening pedigree* notes the proband, siblings, half-siblings, parents, offspring, and other affected or otherwise relevant relatives, as well as ethnic

TABLE 9-3
Information Sought in the Course of Taking an Adequate Family History

1. Affected family members other than the proband.
2. Range of disorder's manifestations in the individuals recognized as affected.
3. Itemization of at-risk persons for further investigation.
4. Age of parents and grandparents relative to conception of first affected person.
5. Ethnic backgrounds of family members.
6. Consanguinity.
7. Incest; nonpaternity.
8. Presence of nondisease family traits (e.g., clinodactyly, facial appearance, etc.).
9. Presence of other heritable or congenital disorders.
10. Presence of other types of health problems.
11. Presence of twinning.

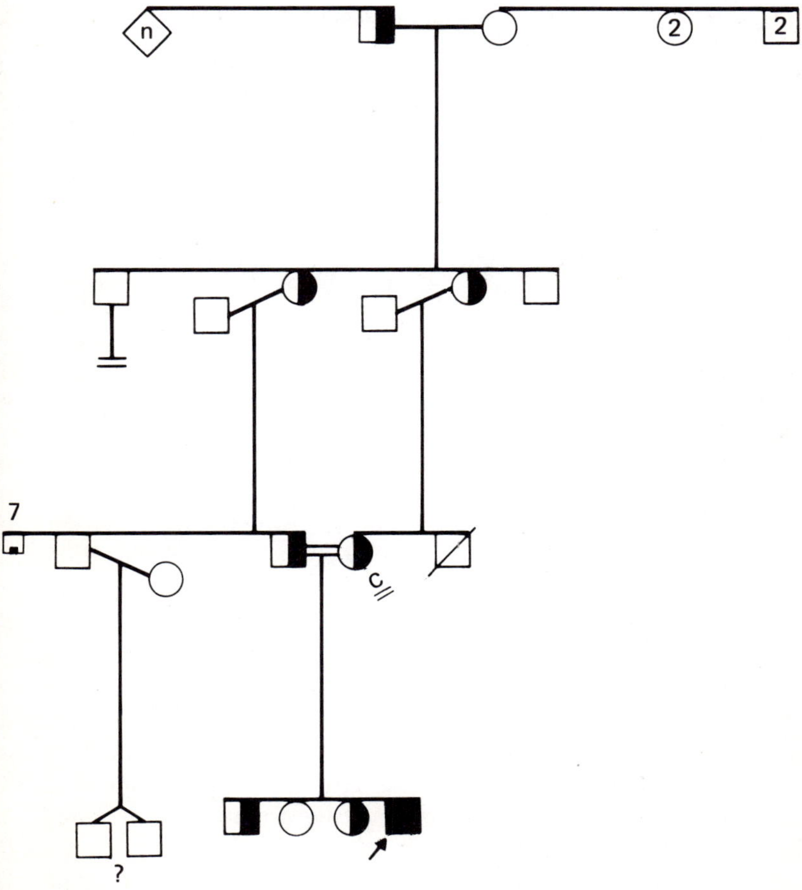

Figure 9–2 A sample pedigree. Autosomal recessive inheritance, with consanguinity, is depicted. The carrier status of the proband's siblings could actually be established only if a carrier test were available; they are included here in any case to emphasize the 2 in 3 chance that a normal sibling to a person with an autosomal recessive disorder will be a carrier.

TABLE 9-4

Guidelines and Rules for Constructing Pedigrees

1. Locate male partner on the left of a mating line, whenever possible.
2. Designate birthdates for proband and each immediate family member, as well as for all affected individuals, if possible; otherwise designate ages.
3. The consultand (marked by C) is the individual who seeks counseling; the proband (marked by an arrow) is the most immediate family member affected. If it is a couple seeking counseling, the consultand is the mother—whether or not she is affected—unless the disorder affects the husband or his blood relatives. An affected young child is usually the proband.
4. Maiden names are written in parentheses after first names. They should be noted for the proband or spouse; mother of the proband; and each affected, carrier, or at-risk married female.
5. Married names should be noted for each married female who is affected, is a carrier, or is at-risk.
6. Sets of siblings, stillborns, and abortions should be arranged in chronological order, with the oldest on the left.
7. Causes of death and/or health problems should be noted for the various family members below or above their respective symbols.

Note: These standard rules, plus the symbols depicted in Fig. 9-1, are the basis for constructing pedigrees as shown in Fig. 9-2.

backgrounds and whether there might be consanguinity or advanced paternal ages. This preliminary pedigree is a starting place for a more detailed examination of the family history; it also gives some indication of the role the disorder has played or will play in the family's domestic life. For example, is the disorder a repetition or a unique phenomenon? Is it a source of strife, as indicated for example by a parent's comments, "It can't have come from *my* side of the family"? The *complete pedigree* deals with at least all known second- and third-degree relatives to the proband, plus more if there are any positive or suspicious findings along the way. Important additional information is often obtained when the complete pedigree is taken. It serves to: 1. define, or at least suggest, the patterns of inheritance if the findings are positive; 2. clarify the mode or modes of manifesting the disorder among those affected or at risk; 3. identify those persons thought to be affected or at-risk, so that they can be evaluated and/or informed of the recurrence risks. This last application of pedigrees, as instruments for "prospective genetic counseling," is assuming increasing importance in clinical genetics (2,3).

The application of computer technology to pedigree data (4) allows for more sophisticated analysis and for rapid access to affected and at-risk family members at some later date when improved therapy and/or diagnostic measures become available. More and more genetics programs are deriving their pedigree data in a manner that facilitates computer processing.

Maternal–Fetal Histories

The *conception history* overlaps with the family history in that prior pregnancies, abortions, miscarriages, stillbirths, and live births are important to both. Periods of apparent infertility, duration and types of contraception, environmental exposures, and personal health histories of the consulting couple are important in any genetic evaluation. Infertility, multiple miscarriages, or stillbirths may have been the indication for genetic referral in the first place. Under those circumstances the aim is to use the derived data and formulations prospectively to facilitate completion of a normal pregnancy with healthy offspring. When there is already an affected infant or child, the evaluation is directed first retrospectively to account for the presenting condition, and thence prospectively to advise about recurrence, or even preventive measures. The point is to determine as accurately as possible whether any environmental or parental factors (e.g., ages, maternal PKU) can account for the disorder, or at least suggest contributing factors. Points of concern include: maternal and paternal ages and health; exposure to environmental hazards through place of residence, occupation or hobby; medications; drug and alcohol habits; discontinuance of birth control pills; removal of an intrauterine device; out-of-wedlock or undesired pregnancy with the possible surreptitious use of an abortifacient. In practice, answers to these questions rarely resolve the issue on their own, but they are a necessary context for further investigative efforts.

The *pregnancy history* itself is obviously significant. First, was there exposure to environmental agents, planned or accidental? Second, what was the pregnancy like from the mother's standpoint—nausea, vomiting, weight gain, high blood pressure, threatened abortion, aggravated diabetes mellitus, etc.? Third, what was the pregnancy like as judged from the fetus's standpoint—onset and intensity of fetal movements, oligohydramnios, etc.? Fourth, what about possible maternal–fetal interaction—ABO or Rh incompatibility, etc.? For the majority of genetic and congenital disorders there are no constant clinical signs during gestation, although diabetes mellitus, severe anorexia, oligohydramnios, and threatened abortion may portend problems. Any radiation exposure, drug ingestion, or infection may be important, but the predictable problems they can cause are limited to relatively few specific agents (see Chapter 7). A nonspecific viral illness and subsequently decreased fetal movement may be a major factor in diagnosing self-limited congenital arthrogryposis (5,6).

A normal gestational history is compatible with any genetic or congenital disorder. Positive histories can help sort out some obscure or equivocal diagnoses.

The *delivery history* is especially important. When did the membranes rupture; how long was labor; how near term was delivery; what type of delivery occurred (vaginal versus cesarean section); where did delivery take place; what type of presentation was it (e.g., breech, vertex, forehead); what kind of anesthetic or other medication was used, etc.? Were there fetal heart tones throughout; was the umbilical cord around the neck; what was the Apgar score; were the weight, length, and head circumference average for gestational age; were any deformities or functional deficiencies immediately obvious; etc.? The main point of these and other queries is to determine, if possible, whether the birth process in any way contributed to immediate or subsequent problems. If there were no anomalies and no delivery problems, then the inquiry would focus on family, developmental, and personal histories. If anomalies were present, specific disorders and etiologies could be considered in an orderly fashion (see below). If anomalies were obvious at birth, were they the primary cause of the proband's delivery problems? For example, delivery may have been difficult because of intrauterine hydrocephalus.

For any given problem, a negative birth history means that a cause operating before and/or after birth must be sought. If congenital anomalies are present, an intrauterine insult (teratogenic, chromosomal, heritable) must be considered. If no anomalies are apparent at birth, teratogenesis is not excluded (e.g., deafness from congenital rubella), but heritable and perinatal problems must be considered first. And, of course, major congenital malformations are sufficiently common that they may be merely coincidental to other problems. In any case, intelligent evaluation of common problems such as failure to thrive, developmental delay, and frank mental retardation can only take place in the context of what is accurately known about the circumstances of birth.

Postnatal Histories

The *neonatal history* may be an extension of the delivery history if there were birth problems. Or problems may develop hours or days after delivery. In either case, one must ask whether there was jaundice, seizures, hypoglycemia, poor feeding, infections, etc.

Documentation of the presence or absence of a normal interval after birth during the neonatal period is directly relevant in evaluating problems with possible neonatal onset.

The *developmental history* involves retrospective assessment of a child's psychomotor development up to the time of the history-taking in order to: determine whether there is retardation in any sphere; and, if present, estimate the time of onset of developmental delay and whether there were any associated problems, such as high fever, cyanosis, seizures, meningitis, or trauma. Formal and informal assessment methods are well known and will not be reviewed here (7). The important considerations are whether the child's current status had any neonatal indicators or correlates (e.g., profound hypotonia, hypoglycemia), whether there was developmental delay from the beginning or only after a period of normal development, and whether the status is arrested or regressing. The point at which the child's development can first be distinguished from normal is of special interest. If distinct since birth, the cause is presumed prenatal (congenital) or perinatal. If the difference shows up later, then a postnatal factor, perhaps precipitating a heritable biochemical disorder, is likely. In these assessments however, it is always important to keep in mind family likenesses; comparison to siblings and other family members as well as to the general population is in order.

The details of the *social history* of the family may provide clues to the cause of either prenatal, perinatal, or postnatal problems. For example, a poor family living in close quarters may be the setting for incest, poor obstetrical care, delivery at home or in an automobile, perinatal infection, poor nutrition, maternal deprivation, child abuse, or education neglect. Similarly, an alcoholic mother may raise suspicions of prenatal problems (fetal alcohol syndrome), social and nutritional neglect, or child abuse.

The *medical history*, including birth measurements and data from periodic examinations since birth, can also help estimate the time of onset of the affected individual's problems. Thus, intrauterine growth retardation suggests an etiology and pathogenesis quite distinct from what is indicated by growth delay beginning months or years after birth. Head size is a particularly important parameter in this regard. Another set of considerations is the onset and progression of organ enlargement. For instance, a child who is apparently normal until age 18 months and whose psychomotor development is then arrested or begins to regress with simultaneous progressive enlargement of liver and spleen, may have a storage disease.

Despite its importance, history-taking is often overlooked in discussions of genetic evaluation, which largely deal with the many available types of cytogenetic, biochemical, and other laboratory modes of assessment. Yet, in fact, because genetic laboratory resources are so limited, the specificity of each test so great, and their total number so large, they can be used only when guided by findings from the histories and physical examination. The histories also provide a general context for appreciating the role of the disorder in the affected person's life.

Physical Examination of Patient and Family Members

Five related concepts underlie the physical examination in a genetic evaluation: attention to minute details, subtlety, timing, patterns, and precise terminology. By attention to gross and subtle physical findings described in minute detail with precise terminology, specific patterns and nonspecific combinations of abnormalities can be appreciated which point to either the timing of the insult and/or a particular diagnosis. Facility in "syndrome recognition" is part of a clinical geneticist's expertise. But all clinicians must recognize potential syndromes or physical findings with special significance in order to bring patients to the geneticist's attention.

From a genetic viewpoint, there are three basic considerations regarding physical findings: 1. Is there essentially only one or clearly more than one organ system involved? 2. Does the abnormality have a prenatal onset (congenital anomaly)? 3. Can findings in other family members give meaning to findings in the proband? That is, is it a nonspecific family trait, or is there partial expression of a disorder in each of several members so that the disorder becomes clear only when a composite is made?

Table 9-5 lists features of special importance in a physical examination for genetic evaluation. Physical features are usefully characterized according to whether they are congenital and as to which organ or type of organ is involved. This list is not meant to be all-inclusive; however, the pattern of positive and negative findings can be very helpful in suggesting the type of disorder, its time of onset, and the etiology. Abnormalities present at birth are of two major types: 1. literal anomalies, representing an aberration in development of a gross morphologic nature (e.g., limb-reduction defects); and 2. disruptions of tissue or cellular function not different in kind from abnormalities present after birth (e.g., ichthyosis, chondrodystrophy). That is, patients with congenital anomalies *per se* are to be distinguished from those whose disorders merely have a prenatal onset. This distinction, plus historical data, will indicate what types of tests are needed.

When disorders present in infancy, a critical consideration is whether the child has congenital anomalies. Table 9-6 shows how

TABLE 9-5

Points to Be Considered in the Course of a Genetic Evaluation Physical Examination[1]

I. Congenital malformations
   A. Single organ system or tissue
   B. Multisystem
II. Central nervous system involvement
   A. Morphologic abnormalities
   B. Mental retardation, developmental delay
   C. Spasticity or hypotonia
   D. Other aberrant functions (e.g., seizures, insensitivity to pain, etc.)
III. Genital, gonadal involvement
IV. Visceromegaly
V. Bone dysplasia
VI. Ocular involvement
VII. Dermatoglyphic abnormalities
VIII. Special measurements

[1]The two key points are whether some or all of the findings have a prenatal onset and whether there is multisystem involvement.

TABLE 9-6

Most Likely Etiologies for Infantile Disorders as a Function of How Many Anomalies Are Involved[1]

Anomalies	*Chromo-somal*	*Single-gene*	*Polygenic*	*Develop-mental*	*Environ-mental*	*Acquired*
None	0	+	0	+	0	+
One	0	+	+	+	+	0
Two or more	+	+	0	+	+	0

[1] These are estimates for deciding on initial evaluation procedures. Note that acquired disorders do not have congenital anomalies as features, that polygenic disorders entail single anomalies, and that chromosome disorders usually have multiple anomalies. Single-gene and developmental disorders are compatible with the presence or absence of anomalies. Conversely, multiple anomalies require a full consideration of all nonacquired etiologies. + = Reasonably likely. 0 = less likely, or unlikely. The etiologic categories are as discussed in Chapter 2.

this question relates to the genetic evaluation. If *no anomalies* are present, a single-gene, developmental, or acquired disorder is most likely. An acquired disorder may be associated with a history of specific insult, such as perinatal anoxia, hypoglycemia, or meningitis. A single-gene disorder may be associated with an informative family history. A developmental disorder has neither association. If only *one anomaly* (or one type of anomaly) is present, the major considerations are single-gene, polygenic, environmental, and developmental disorders. The polygenic anomalies are of certain specific types (see Chapter 5), and only one directly involves the extremities (club foot). Single-gene disorders usually conform to previous descriptions and/or may have an informative family history. Single anomalies may be caused by environmental agents (e.g., cleft palate and diphenylhydantoin), although this is unusual. A developmental diagnosis (subcategory C1, Chapter 2) for single-organ anomalies is not merely based on exclusion, but on careful description and published reports; many limb malformations, such as congenital absence of a tibia, fall into this category. *Multisystem malformations* (two or more anomalies) indicate an early, diffuse intrauterine insult, perhaps as a result of a chromosome abnormality, single-gene mutation, or teratogen. Chromosome studies and screening tests for infectious teratogens should thus be initiated immediately, and detailed histories obtained regarding family members and pregnancy exposures. Developmental disorders should be included for consideration here, based on established precedent (e.g., amniotic bands syndrome).

Note that single-gene and developmental disorders are to be considered for all three groups in Table 9-6, i.e., no matter how many anomalies are involved. On the one hand, this indicates that a heritable basis should be sought for all infantile disorders not documented as acquired disorders; on the other, it indicates that an etiology often escapes identification. With multiple anomalies, chromosomal and teratogenic etiologies must be given special consideration if the central nervous system is involved. None of these statements are firm rules; rather, they are guidelines for initiating diagnostic evaluations. They do not apply to circumstances which require unusual explanations, like the coincidental occurrence of two common malformations or the notion of developmental field defects (Chapter 7).

Documentation of central nervous system involvement in an individual presenting for genetic evaluation can be especially important, since it portends substantial morbidity and high burden. More than one-third of genetic referrals show some neurologic deficit. A demonstrated or presumed congenital aberration, especially of a morphologic nature, associated with delayed psychomotor development always warrants a chromosome analysis unless another etiology (e.g., teratogen) has been documented. Virtually every type of neurologic problem can have a single-gene etiology in certain instances.

All disorders involving anomalous genitalia, intersexuality, or hypogonadism must be evaluated from a genetic viewpoint, beginning with a chromosome analysis. Many disorders of each type are known to have a heritable basis.

Visceromegaly suggests a heritable disorder of a biochemical type, especially if clinical problems develop after a period of normalcy. The same is true of ocular abnormalities. Bone dysplasias—aberrations of bone growth with distortion of the epiphyses, metaphyses, or diaphyses—should be presumed heritable until proven otherwise (8,9). Organ enlargement and bone involvement may also suggest an infectious teratogen.

Geneticists closely examine the palms, soles, and fingers with regard to their patterns of creases and dermal ridges (e.g., finger prints). This *dermatoglyphic analysis* can be significant, since deviations from the normal may either suggest the presence of an otherwise subtle disorder or indicate the need for a chromosome analysis when a person is already known to have a disorder. In addition, since the dermal ridges are formed by 16 weeks' gestation, their aberration documents the prenatal onset of the disorder. Care must be taken, however, to keep in mind certain normal variants which may be familial. The details of dermatoglyphic analysis can be found in many places (10,11). Here, it is enough to note that the most useful parameters are transverse palmar (simian) creases, distal displacement of the palmar axial triradius, single-flexion creases on the fifth fingers, five or more arch patterns among the ten fingertips, and an arch tibial pattern in the hallucal region (base of the great toe). These patterns are schematized in Fig. 9-3. An example of their application would be the case of a ten-year-old boy who is apparently normal except for

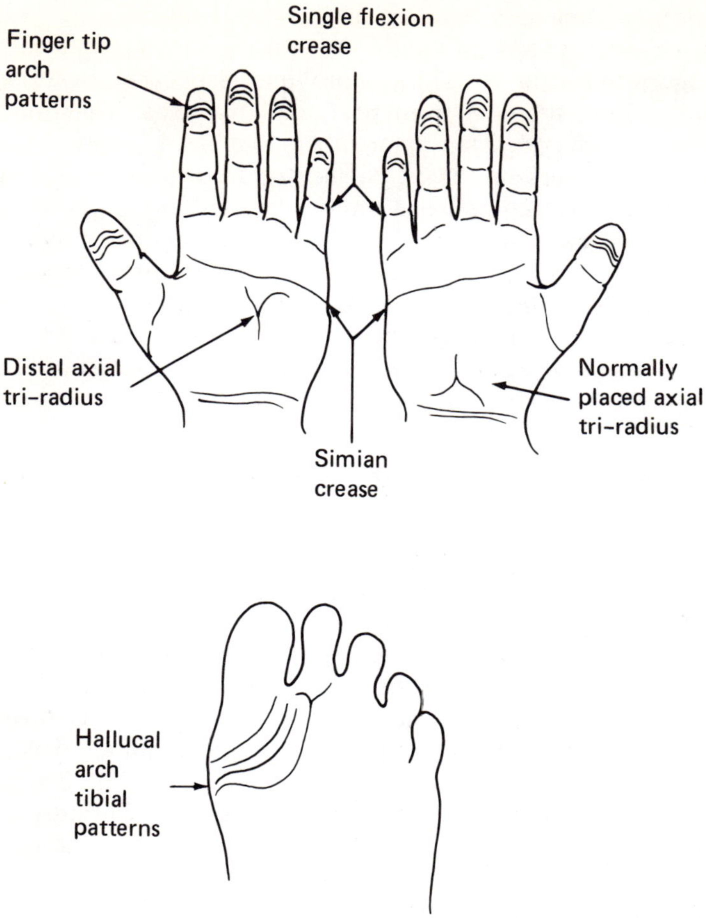

Figure 9–3 Schematized depiction of the most readily identified dermatoglyphic aberrations of the fingers, palms, and soles.

small testes and distal palmar axial triradii—a chromosome analysis is warranted to determine whether an extra X-chromosome is present (i.e., 47,XXY). A child exhibiting mental retardation and ten fingertip arches should also have a chromosome analysis. Patterns of scalp hair growth can have similar diagnostic use (12). In addition to length, weight, and head circumference, other body parts can be measured and deviations from the average taken to indicate a possible genetic abnormality (13). The most helpful measurements are arm span, upper/lower segment ratios, ear size, distance between the inner corners of the eyes, distance between the outer corners of the eyes, distance between the midpoints of the pupils, and length of the palpebral fissures.

Thus, in summary, the physical examination has a similar

function in genetic evaluation as in other medical settings, but with more emphasis on subtle deviations from normal and on timing the onset of the clinical problem, especially in terms of distinguishing prenatal from postnatal problems. The frequent need to examine other family members to make a diagnosis in the proband also bears emphasis. The central role of physical diagnosis in distinguishing between hundreds of different, overlapping dysmorphology syndromes emphasizes the need for familiarity with a wide-ranging literature and for recording the patient's features by photograph. The photography allows for comparisons over time, enhances verbal descriptions, and permits clinicians not in contact with the patient to participate more meaningfully in diagnosis.

Laboratory Components of a Genetic Evaluation

Three basic types of laboratory data are used in the course of a genetic evaluation: 1. Data from routinely available tests not limited to use where a genetic diagnosis is in question (e.g., complete blood count, serum electrolytes, urinary excretion of steroid hormones). 2. Data from specialized tests not limited to use where a genetic diagnosis is in question (e.g., liver biopsy, computerized x-ray examinations). 3. Data from tests directed specifically at establishing or clarifying a genetic diagnosis (e.g., chromosome analysis, enzyme assays, amniocentesis). The third type may be the responsibility of a genetics laboratory (e.g., chromosome analysis, most enzyme assays) or of another specialty lab (e.g., hematology).

Laboratory tests can also be classified from an operational perspective: 1. Incidental screening: tests that incidentally suggest a genetic disorder (i.e., a blood smear showing sickle cells or spherocytes). 2. Genetic screening: tests that are used for preliminary screening of genetic disorders (i.e., Berry spot test for urinary mucopolysaccharides). 3. Definitive tests: tests that specify the genetic diagnosis (e.g., karyotyping, enzyme assay). The genetic significance of incidental screening tests and the uses of the genetic screening tests, chromosome analysis, and commonly available enzyme assays should be known to all primary-care physicians. The main purpose of this section is to clarify the kinds of tests relied upon in a genetic evaluation.

In the course of a genetic evaluation, the clinical geneticist must not only be able to suggest the appropriate tests, but must also make them available, directly or indirectly, and interpret or assist in interpreting the results. Considering the many different types of tests potentially relevant to a given case, their typically limited availability, and their substantial cost or investigative nature or mutual exclusiveness, knowing when to order which test is a key element in a genetic evaluation. The geneticist's use of available laboratory resources on behalf of a given patient brings us back to the *nidus* model of his role in collecting and processing information. The laboratory information is used to specify or exclude a certain etiology, to identify associated features of the disorder, and to extend our understanding of genetic disease. Since genetic aberrations—whether chromosomal or mutational—ultimately entail an intracellular defect, genetic tests are aimed at demonstrating abnormalities of cellular function, directly or indirectly. The "nongenetic" tests, such as those which demonstrate infectious teratogens, are aimed at another level. The closer the genetic test is to measuring the genetic material (e.g., chromosome analysis, enzyme assay), the more specific is its significance for the patient, both in terms of recurrence risks and management.

Table 9-7 lists the laboratory tests (including x-rays) that are commonly ordered or urged by the clinical geneticist during genetic evaluations. Although the list is relatively comprehensive, it is neither exhaustive nor uniformly detailed. On the one hand, it indicates the range and depth of the differential diagnoses involved and the problematic nature of some of them, while on the other, it reemphasizes the nidus concept of clinical genetics. A genetic evaluation involves the collection of many types of complex data, often from distant places and over long periods of time, and the geneticist is responsible for this compilation. Rather than merely recommending that something be done, the geneticist ensures that it is done and that the acquired data are meaningfully utilized.

A very common diagnostic problem is the infant or child with *multiple congenital anomalies*, born of an unremarkable pregnancy, whose family history is unrevealing. If a specific chromosomal disorder is apparent (e.g., Down syndrome), a chromosome analysis must be done and associated features sought by such methods as intravenous pyelography (IVP) and electrocardiography. If no specific diagnosis is apparent, and especially if the central nervous system is involved, chromosome analysis and TORCH screen are warranted, plus the search for associated features. If the anomalies are restricted to the genitalia and gender is ambiguous, an immediate search for adrenal hyperplasia may be warranted, plus a buccal smear for X- and Y-chromatin and a chromosome analysis. Radiographic contrast studies of the internal genitalia should be performed, as well as an IVP, and perhaps cystoscopy. A prompt, definitive evaluation of ambiguous genitalia can lead to life-saving measures, clarify gender for optimal child-rearing, provide a basis for appropriate surgery, and define the long-term expectations and clinical course.

If *no anomalies* are involved and the problem is developmental

TABLE 9-7
Itemization of Laboratory Studies Frequently Ordered or Suggested Directly by the Geneticist During Genetic Evaluations. (Numbers in Parentheses Refer to Other Relevant Chapters in This Book.)

1. Karyotyping (3)
2. Sex chromatin analysis (3)
3. Urine metabolic screens or quantitative analysis (6)
   a. Sugars
   b. Organic acids
   c. Amino acids
   d. Mucopolysaccharides
   e. Uric acid
   f. Other
4. Enzyme assays (6)
5. Amniocentesis (8)
6. Hemoglobin electrophoresis
7. Bone marrow biopsy for storage cells and hematologic analyses (6)
8. Other tissue biopsies—for histology, ultrastructure, and biochemical assays.
   a. Muscle
   b. Rectal
   c. Skin
   d. Liver
   e. Renal
   f. Nerve
   g. Gonadal
9. Electrophysiologic studies
   a. Electrocardiography
   b. Electroencephalography
   c. Electromyography
   d. Nerve conductions
   e. Electroretinography
10. Audiograms
11. Serum and red blood cell antibody/antigen profiles
12. Human leucocyte antigen (HLA) profiles
13. Serum immunoelectrophoresis
14. Radiographic studies
    a. Intravenous pyelogram
    b. Genitogram
    c. Skeletal—partial or complete survey
    d. Bone age
    e. Computerized axial tomography
    f. Hysterosalpingogram
    g. Other
15. Ultrasound (sonography)
    a. Prenatal diagnosis
    b. Cardiac
    c. Renal
    d. Other
16. Red blood cell osmotic fragility
17. Blood coagulation factors levels
18. Serum levels of muscle enzymes
19. Liver function tests
20. Renal function tests
21. Glucose tolerance test
22. Serum levels of cholesterol and triglyceride, or lipoprotein electrophoresis

23. Blood metabolite levels
    a. Uric acid
    b. Ceruloplasmin
    c. Phosphorus, calcium, zinc, copper, etc.
    d. Amino acids
    e. Serum protein electrophoresis
    f. Other
24. Hormone levels in blood, urine
25. Spinal fluid analysis for specific metabolites (e.g., 5-hydroxyindole acetic acid)
26. Intestinal absorption studies
27. Sperm count and characterization
28. Serologic and microbiologic studies to identify infectious teratogens, e.g., serum TORCH screens (7)

delay, with no period of intervening normalcy, chromosome analysis would have a very low yield. A TORCH screen should be done, but with emphasis on metabolic screening and neurologic evaluation. The neurologic tests might include electromyography, computerized radiotomography of the cranium, and electroencephalography. Metabolic evaluation would include determination of acid–base balance, blood ammonia levels, and urine screen for increased amounts of reducing substances (sugars, etc.), amino acids, and organic acids.

If a period of normal development is followed by *deterioration* of central nervous system function, the metabolic considerations should be paramount. If coarsening of facial features, organomegaly, or ocular or skeletal changes develop, a storage disease or related disorder (e.g., Krabbe disease) should be suspected. Storage disorders are pursued by screening the urine for mucopolysaccharide excretion, performing a skeletal survey, and assaying for a battery of specific enzymes in serum, peripheral white blood cells, and/or cultured skin fibroblasts. If the skeletal x-rays suggest a mucopolysaccharidosis but there is no mucopolysacchariduria, then any of a variety of mucolipidoses or other lysomal storage diseases should be investigated by enzyme assays.

Hypogonadism always raises the possibility of a sex chromosome abnormality, even if there are no other apparent clinical problems. The minimal evaluation includes a buccal smear for X-chromatin and Y-chromatin, and a peripheral blood chromosome analysis. Other tests, such as assays of sex steroid levels in blood and urine and radiography of internal genitalia (e.g., genitogram), may also be appropriate.

The foregoing examples give an idea of how some of the tests listed in Table 9-7 can be used in certain circumstances, depending on the history and physical examination. There are no "genetic tests" to be applied uniformly for all genetic evaluation referrals. In Table 9-8 laboratory tests are arranged in terms of presenting problems with emphasis on impairment of intellectual development and congenital anomalies. These two types of problems probably represent at least two-thirds of genetic referrals (Tables 2-4, 2-5, 2-6).

Linkage analysis is a special kind of genetic laboratory approach. Its purpose is to infer one gene's presence from the demonstrated presence of a second gene. This is done by empirically establishing an association of one gene with another by family studies (14) or somatic-cell culture techniques (15). It can be clinically useful when each of the two loci (e.g., A and B) has more than one allele, so that for any given family the presence of one allele at locus A indicates the likelihood of a specific allele at locus B. That is, one allele serves as a marker for a given allele at a different locus. All genes on the X-chromosome are linked. Linkage groups for autosomes have only recently been established, i.e., between the ABO blood group and the Nail-patella syndrome (ABO/NPS) (16), and between the ABH-secretor locus and myotonic dystrophy (17). For either of these two systems, the demonstration of a given blood group phenotype implies the presence of the mutant gene for the disorder, even before the disorder becomes clinically apparent. This approach will become increasingly useful for pre-onset diagnosis of dominant disorders that begin relatively late in life, such as Huntington chorea. It can even afford prenatal diagnosis of certain disorders, such as myotonic dystrophy (18). Although linkage studies currently have limited clinical application, the situation will change rapidly for the following reasons: 1. Increasing numbers of useful markers are being recognized at the same time as family study methods are improving. 2. Computers have maximized the likelihood of gleaning the needed information from available data on families. 3. Somatic-cell genetic techniques (19) have made it possible to link genetic loci and assign them to given chromosomes even without family studies.

The most important limitation to linkage analysis is the distance between the loci in question, respecting the phenomenon of crossing over in meiosis. The farther apart two loci are, the greater the likelihood of crossing over between them, and, therefore, the greater the chance that the marker allele will be disjoined from the mutant allele being sought. To be clinically useful, the two loci have to be so close that crossing over will not be a large factor. The second limitation is that for each family the precise nature of the marker allele linked to the mutant allele in question must be established, since it can vary from one family to another. With the ABO/NPS system for example, it must be shown for a given family whether the A, B, or O allele is linked with the NPS allele.

TABLE 9-8

Preliminary Laboratory Considerations in the Early Stages of Evaluating Representative Types of Presenting Problems[1]

	Buccal Smear	Karyo type	Metabolic Screen[2]	WBC Fibro-blast Enzyme[3] Assay	Hormone Screen[4]	TORCH Screen	X-ray Skeletal Survey[5]	Cranial CAT X-rays[6]
I. Multiple congenital anomalies								
A. Without CNS (central nervous system) involvement	–	±	–	–	–	?[7]	+	+
B. With CNS involvement	–	+	–	–	–	+	+	+
II. Developmental delay/retardation								
A. With congenital anomalies	–	+	–	–	–	+	+	+
B. Without congenital anomalies								
1. No initial normal period	–	+	+	±	+	+	+	+
2. After initial normal period								
a. With storage features[8]	–	?[9]	+	+	–	?[7]	+	?[10]
b. Without storage features	–	?[9]	+	+	±	?[7]	+	+
III. Skeletal dysplasias	–	–	+	?[11]	+	–	+	–
IV. Multiplo miscarriages	–	+	+	–	+	–	–	–
V. Infertility	–	+	+	–	+	–	–	–
VI. Hypogonadism	+	+	–	–	+	–	+	–
VII. Ambiguous genitalia, intersex	+	+	–	–	+	–	–	–

[1]+ = should be actively considered; – = not likely to be helpful; ± = may occasionally be useful; ? = may be useful under limited circumstances.
[2]As per Table 6-2 and/or bone marrow biopsy to look for storage.
[3]As discussed in Chapter 6, pp. 117–120. WBC = white blood cell.
[4]Blood or urinary levels of various hormones or their metabolites, according to the differential diagnosis.
[5]A complete visualization of the skeleton, including bone age assessment.
[6]CAT = computerized axial tomography.
[7]TORCH agents (see Chapter 7) are not usually implicated in these categories, but under certain circumstances a screen might be warranted, e.g., if death is too soon to evaluate CNS function, or early developmental history is incomplete.
[8]Storage features include such things as organomegaly, heart dysfunction, renal dysfunction, corneal clouding, cataracts, etc.
[9]Karyotyping is usually not warranted if there is an early, normal, developmental period, but if this is not fully clear, chromosome analysis becomes reasonable.
[10]Although primary brain abnormalities are not expected as a part of storage diseases, if early data are equivocal or there are concerns for secondary changes (e.g., hydrocephalus), this test becomes reasonable.
[11]Currently, skeletal dysplasias not part of one of the other categories are not definitely associated with enzyme deficiencies in white blood cells or skin fibroblasts.

The Library in Genetic Evaluation

All clinical specialists are obliged to keep abreast of the literature in their field. For the clinical geneticist the problem has at least four dimensions:

1. The literature is vast and relevant reports appear in many major journals and many different professional meetings. A familiarity with the major journals of each subspecialty is a basic requirement. For example, reports pertinent to genetic counseling for Duchenne muscular dystrophy appear in the subject genetics, neurology, pediatrics, general medicine, surgery, pathology, and many basic sciences.

2. Many of the disorders to be considered in a differential diagnosis may have been the subject of only one previous case report. Clinical geneticists are expected to be familiar with rare conditions, and this requires a heavy reliance on books, journals, and meetings, as well as personal experiences.

3. If the disorder is not known or described, one should publish a report of it for at least two reasons: (a) It may clarify the situation for other cases. (b) Professional responses to the report may in turn clarify certain aspects (e.g., prognosis, heritability) for the case reported.

The first two points—large, diverse literature and rarity of many of the disorders—explain why a genetic evaluation often takes time. At the first visit, details are obtained and then the literature is searched before meaningful counseling can be given at a later date. The third point—apparent uniqueness of the case—explains why genetic evaluations often lead to reporting the case in one form or another.

4. Like other specialists, the clinical geneticist has an obligation to relate pertinent literature directly to a given family. Thus, if new reports change the diagnostic considerations, prognosis, risks, or alternative course of action, the clinical geneticist may be obligated to contact the family again for reevaluation and/or further counseling. This obligation, in turn, indicates a need to maintain an awareness of past as well as present patients and a means for ready access to them.

Collection, Collation, and Formulation of Data

In keeping with the nidus concept of genetic counseling, the initial contact with the patient or family is just one part of the process leading to a final disposition. From experiences in the Colorado-Wyoming Regional Genetic Counseling Program (Chapter 16), we have calculated that on the average a minimum of 7.5

hours is spent on arriving at and effecting a final disposition for a family (Table 9-9). In the vast majority of cases the process is spread out over weeks or months. The time factor has several practical implications: 1. The family and whoever referred it should not expect the desired answer at the initial contact with the geneticist. 2. Staffing, time allotments, record-keeping, and organization of the genetic counseling service must reflect these complexities. 3. It is virtually impossible for each family to be charged according to the time spent on their behalf.

The *collection of data* has two main components. First, data are derived directly from the patient and family by histories and examinations. Second, data are derived from old records, laboratory procedures, and discussing the situation with colleagues in genetics or other disciplines. The need for sophisticated laboratory backup and for expertise in other subspeciality areas underscores the requirement that a genetics clinic be part of a large medical center. The *collation of data* entails their compilation and arrangement, so that the major problem and its various facets are delineated. Based on this collation, the *formulation* involves review of the relevant literature, reliance on the geneticist's acumen, and presentation of the case at clinical conferences held specifically to review patient evaluations. At these conferences the need for additional evaluation is suggested, or a diagnosis is established by concensus and the planned counseling is discussed. The role of photography here is important: photographs of the patient and/or family members allow others to more meaningfully participate in the formulation. From this entire input a *disposition* is derived which is transmitted to the family at the

TABLE 9-9
Operational Components of a Genetic Evaluation, Based on an Average Time Expenditure of 7.5 Hours per Family[1]

	Average Time Spent (hours)	%
1. Data collection—patient/family contact	2.0	26.7
2. Data collection—other (old records, laboratory analyses, consultation, etc.)	2.0	26.7
3. Data collation (initial review)	0.5	6.7
4. Formulation (literature review, patient conferences, synthesis, etc.)	2.0	26.7
5. Disposition, genetic counseling	1.0	13.3
Total	7.5 hours	100%

[1]This table is based on an average time expenditure of 7.5 hours per family. Note that only 40% of the time is actually spent with the patient or family (items 1 and 5). Additional followup sessions, though frequent, are not included here.

second or third clinic visit. Although every case does not require the same amount of time and effort, the general approach is the same. The process is deliberate, time-consuming, and usually complex. It always entails consensus by knowledgeable experts and sensitive, thoughtful transmission of the final conclusions (to be discussed in Chapter 10). The minimum time spent on behalf of a family is 2½ to 3 hours, and not infrequently 15 to 20 hours may be required.

Cited References

1. Jones KL, Smith DW, Harver MAS, et al: Older paternal age and fresh gene mutation: Data on additional disorders. *J Pediatr* 86:84, 1975.

2. Frouhar Z, Spiro R, Lubs ML: The advantage of thorough investigation in families with X-linked diseases. *Am J Hum Genet* 26:32A, 1974.

3. Hathaway H, Lubs M, Kimberling W, et al: Carrier detection in classical hemophilia. *Pediatrics* 57:251, 1976.

4. Kimberling WJ: Computers and gene localization. in Wright SW, Crandall BF, Boyer L (eds): *Perspectives in Cytogenetics*. Springfield, IL, Charles C. Thomas, 1972, pp 131-147.

5. Hall JG: personal communication.

6. Hall JG, Greene G, Shinkoskey S, et al: Arthrogryposis—clinical and genetic heterogeneity. *Clin Res* 24:171A, 1976.

7. Frankenburg WK, Goldstein AD, Camp BW: The revised Denver Developmental Screening Test: Its accuracy as a screening test. *J Pediatr* 79:988, 1971.

8. Spranger JW, Langer LO, Wiedemann HR: *Bone Dysplasias—An Atlas of Constitutional Disorders of Skeletal Development*. Philadelphia, WB Saunders, 1974.

9. Rimoin DL: The chondrodystrophies. *Adv Hum Genet* 5:1, 1975.

10. Shiono H, Kadowaki J: Dermatoglyphics of congenital abnormalities without chromosomal aberrations; A review of clinical applications, *Clin Pediatr* 14:1003, 1975.

11. Mulvihill JJ, Smith DW: The genetics of dermatoglyphics. *J Pediatr* 75:579, 1969.

12. Smith DW, Gong BT: Scalp hair patterning as a clue to early fetal brain development. *J Pediatr* 83:374, 1973.

13. Feingold M: Normal values for selected physical parameters: An aid to syndrome identification. *Birth Defects Original Article Series*, vol X, No 13, 1974.

14. Stern C: Linkage, crossing over and chromosome mapping. in *Principles of Human Genetics* (3rd ed). WH Freeman, San Francisco, 1973, pp 334-374.

15. Ruddle FH, Kucherlapati RS: Hybrid cells and human genes. *Sci Am* 231:36, 1974.

16. Renwick JH, Schulye J: Male and female recombination fractions for the nail patella:ABO linkage in man. *Ann Hum Genet* 28:379, 1965.

17. Harper PS, Rivas ML, Hutchinson B, et al: Genetic linkage confirmed between the locus for myotonic dystrophy and the ABH-secretion and Lutheran blood group loci. *Am J Hum Genet* 24:310, 1972.

18. Schrott HG, Omenn GS: Myotonic dystrophy: Opportunities for prenatal prediction. *Neurology* 25:789, 1975.

19. Davidson RL: Gene expression in somatic cell hybrids. *Ann Rev Genet* 8:195, 1974.

General References

A. Bergsma D: *Birth Defects—Atlas and Compendium.* Baltimore, Williams and Wilkins, 1973.

B. Fraser FC: Genetic Counseling. *Am J Hum Genet* 26:636, 1974.

C. Gustavson KH, Jorulf H: Recurrence risks in a consecutive series of congenitally malformed children dying in the perinatal period. *Clin Genet* 9:307, 1976.

D. Kaveggia EG, Durkin MV, Pendleton E, et al: Diagnostic/genetic studies on 1,224 patients with severe mental retardation. *Third Congress Intl Assoc Sci Study Ment Defic,* The Hague, 1973, pp 82-93.

E. McKusick VA: *Mendelian Inheritance in Man—Catalogs on Autosomal Dominant, Autosomal Recessive, and X-Linked Phenotypes* (4th ed). Baltimore, Johns Hopkins University Press, 1975.

F. Poznanski AK: *The Hand in Radiologic Diagnosis.* Philadelphia, WB Saunders, 1974.

G. Schanman B, Alter M: *Dermatoglyphics in Medical Disorders.* New York, Springer-Verlag, 1976.

H. Smith DW, Simons FER: Rational diagnostic evaluation of the child with mental deficiency. *Am J Dis Child* 129:1285, 1975.

I. Smith DW: *Recognizable Patterns of Human Malformations* (2nd ed). Philadelphia, WB Saunders, 1976.

J. Warkany J: *Congenital Malformations—Notes and Comments.* Chicago, Year Book Medical Publishers, 1971.

K. Wood JW: The pediatrician as genetic counselor. *Pediat Clin No Am* 21:401, 1974.

CHAPTER TEN

TECHNIQUES OF GENETIC COUNSELING

Introduction

Whatever else it may be, genetic counseling is a form of medical care. This fact has two practical consequences: 1. Since the activities that characterize genetic counseling are not different in nature or application from those of other medical disciplines, the professional, ethical, and legal responsibilities inherent in medicine are also inherent in genetic counseling. 2. Genetic counseling cannot be treated as an entity unto itself for the purpose of measuring its efficacy and impact. These points are important because there is a tendency to look upon genetic counseling as if it were far removed from the day-to-day practice of medicine.

The separation of genetic evaluation from genetic counseling in this text belies the continuum that joins these two elements of clinical genetics. It is the net effect of the entire counseling effort, from the initial contact through data collection and formulation to the time they receive a final written report, that determines its value to a family. The genetic counseling *per se* is part of a broader clinical involvement with the family.

What is Genetic Counseling?

Genetic counseling can be viewed as a finished product (i.e., "the family has received genetic counseling"), or as a process—that of informing a person or family of various facts and considerations. In this section we will be concerned with genetic counseling as a process. This process may be considered to comprise only *deliberate* efforts by a clinician to provide counseling or both deliberate and *incidental* activities whose net effect is the counseling. We will take the latter view.

As information is gathered in the course of a genetic evaluation, it is also delivered to the family, deliberately or incidentally. These early stages of transmitting information to the family are no less a part of the counseling effort than the final formulation. Thus it makes sense to distinguish two main stages in genetic counseling (Fig. 10-1): preliminary and definitive. *Preliminary genetic counseling* involves deliberate and incidental information delivery before a final disposition is reached. *Definitive genetic counseling* is the deliberate delivery of the formulated information in a final disposition. The preliminary counseling may be given by any involved clinician but is usually done by the referring (primary) physician or a geneticist. The definitive counseling may be done by the primary physician or the geneticist. This emphasizes the role that can be played by nongeneticists in genetic counseling, in contrast to some definitions of the process that virtually exclude nongeneticists and incidental information delivery (1).

Figure 10-2 diagrams the levels of interaction between a primary physician and a geneticist. The overall point is that there are many patterns of interaction between these two health care providers, and, further, that genetic health care is not the private domain of geneticists. As used here, the term "geneticist" may refer to a single person or to the whole team that participates in the genetic evaluation and counseling. "*Preliminary genetic counseling*" refers to both the evaluation and the initial discussions with the family or patient (i.e., what tests are to be done and

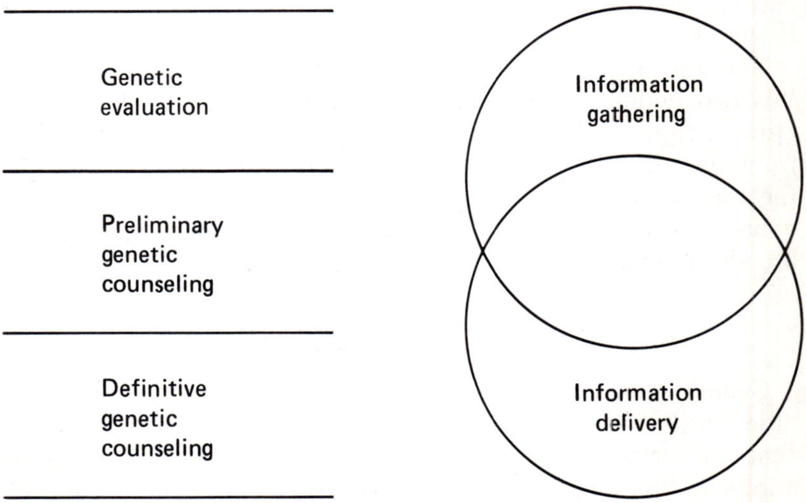

Figure 10–1 Preliminary genetic counseling: a product of the overlap of information gathering and delivery efforts. Preliminary genetic counseling, as portrayed here, is a result of certain information being transmitted by the counselor as other information is being gathered. This depiction emphasizes the need for nongeneticist clinicians to have some genetic background and skills. It also emphasizes that evaluations of genetic counseling as a mode of health care have to consider more than the definitive counseling component.

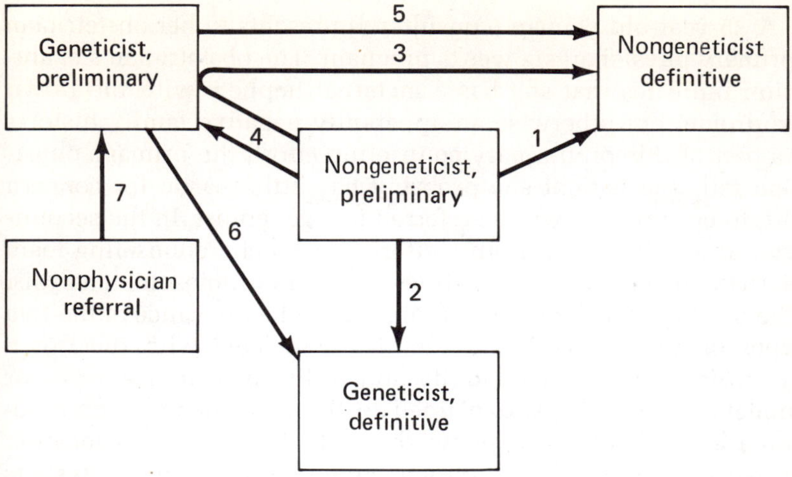

Figure 10–2 Interrelationships of preliminary and definitive genetic counseling as a function of geneticist and nongeneticist responsibilities. This diagram indicates that usually (and ideally) genetic counseling begins with nongeneticist physicians. They may then either directly provide definitive counseling or refer the family to a geneticist who in turn provides the definitive counseling or refers the family back to the primary physician for it. *Arrow 1* indicates that the nongeneticist (primary) physician who has proceeded with the preliminary counseling directly provides the definitive counseling. *Arrow 2* indicates that the primary physician has completed the preliminary counseling and refers the family for definitive counseling. *Arrow 3* indicates that a *minor* second-level preliminary counseling contribution is made by the geneticist (e.g., corroboration of diagnosis) before the primary physician does the definitive counseling. *Arrow 4* indicates that a *major* second-level preliminary counseling contribution is sometimes made by the geneticist (e.g., establishment of diagnosis, family studies, etc.) before any definitive counseling is provided. *Arrow 5* indicates that the geneticist, after a major second-level preliminary counseling effort, refers the family (back) to the primary physician for definitive counseling. *Arrow 6* indicates that the geneticist, after completing the preliminary counseling, proceeds directly to providing the definitive counseling. *Arrow 7* indicates that some families are not referred by physicians, and the geneticist thereby assumes responsibility of all the preliminary counseling; definitive counseling could then be provided either by a nongeneticist physician through referral (arrow 5), or directly by the geneticist (arrow 6). There are thus many ways that nongeneticist primary physicians and geneticists may interact to effect genetic counseling for a family.

why; what the various diagnostic considerations are). A second level preliminary counseling may involve confirmatory evaluation steps, final clarification of details, indication to the family that the information has been formulated, and perhaps even discussion of some of the conclusions. "*Definitive genetic counseling*" refers to the formal session with the family in which the conclusions and the information they are based on are explained in detail, questions are answered, arrangements made for followup, and a written record of the diagnostic measures and conclusions provided to the family. The following example will illustrate this process.

A 25-year-old woman (consultand) presents to her obstetrician (primary physician) six weeks pregnant. The obstetrician's evaluation indicates that she has a maternal nephew with the Down syndrome, but otherwise an apparently negative family history. As part of this preliminary counseling effort, the primary physician tells the patient she probably has little reason for concern but, to be sure, she will be referred to a geneticist. In the second-level preliminary counseling effort, the genetic counseling team learns that the affected child never had a chromosome analysis. The need for checking his chromosomes, the relevance of the two types of chromosome abnormalities associated with the Down syndrome, and the availability and utility of amniocentesis for prenatal diagnosis are explained to the consultand. A chromosome analysis of the nephew shows that a 14/21 translocation accounts for the Down syndrome, and a chromosome analysis is therefore performed on the consultand. Her chromosome constitution involves a balanced carrier state for the translocation. In the definitive genetic counseling session, she is informed of her 15-20% risk of having an affected child in each pregnancy and the details of amniocentesis are discussed. Since she elects to have amniocentesis, the counseling followup has special significance. The results indicate a fetus with the unbalanced chromosome constitution and she elects to terminate the pregnancy. In another followup session, she is reminded that amniocentesis would be appropriate in each subsequent pregnancy and is given a letter containing that information.

The division of labor in genetic counseling between primary practitioners and geneticists is shown in Table 10-1. Although the percentages may be off by a few points, the proportions are probably about right, and this underscores once again the overlap in the activities of primary physicians and genetic counselors.

At each level of genetic counseling, the following principles can serve as guidelines for implementing this type of health care and, ultimately, for evaluating it (Table 10-2).

1. An *accurate diagnosis* is obviously essential for rational counseling.

2. The central element in genetic counseling is *communication*—the need to inform individuals or couples about the risks and alternatives in a way that makes the information useful to them. This means using nontechnical language, setting an unhurried pace, appreciating psychological factors, scheduling at least one followup visit, and providing a written counseling report.

3. The information and alternatives are presented in as *nondirective* a manner as possible. It is most important that those who receive counseling be able to choose a course of action on their own, without being pushed or manipulated.

4. The counselor must be sensitive and adjust to the *psychological and emotional turmoil* that can be expected when any family learns that a genetic or congenital disorder has been diagnosed or is even being considered.

5. In many respects the *family unit is the focus of health care*, and the counselor's concern extends beyond the affected individual (see Chapter 2).

6. More and more, the consequences of genetic counseling include *enhanced management* of the patient and family.

7. The genetic counseling process is *not circumscribed* or removed from other aspects of health care. Many health care providers cooperate during the genetic evaluation and counseling and nongeneticist physicians provide all levels of counseling in some cases.

8. The process of genetic counseling is *time consuming*. A fair amount of time (days to months) usually elapses between initial contact and final disposition. As noted in Chapter 9, an average of 7.5 hours is devoted to each family by a clinical genetics unit. The counseling sessions themselves are time consuming; 1½ hours is a minimum, 2 hours is the usual length, and occasionally up to 3 hours are spent in the initial and followup sessions.

9. A *team approach* allows for both efficient use of each mem-

TABLE 10-1
Relative Proportions (%) of Preliminary and Definitive Genetic Counseling Done by Geneticists and Nongeneticist Primary Physicians

	A, Primary Physician	*B, Geneticist*	*Total*
I. Preliminary genetic counseling	75	25	100%
II. Definitive genetic counseling	25	75	100%
Total	100%	100%	

Notes: 1. (column A) Of the genetic counseling that a primary physician does, about 75% is probably at the preliminary level and 25% at the definitive level.
2. (column B) Of the genetic counseling done by a geneticist, probably about 25% is at the preliminary level and 75% at the definitive level.
3. (row I) For preliminary genetic counseling, in at least 75% of cases the primary physician is probably involved, and in 25% of cases a geneticist may be solely responsible.
4. (row II) For definitive genetic counseling the primary physician is probably responsible for 25%, and the geneticist for 75% of the cases.
5. (column A, rows I and II) Primary physicians are probably responsible for definitive counseling of one-third or fewer of the families they see for preliminary counseling.
6. (column B, rows I and II) Geneticists probably are solely responsible for both preliminary and definitive counseling in about one-third of their families. The percentage figures are estimations by the author from genetics clinic records at two major medical centers (Denver and Milwaukee) and the Colorado-Wyoming Regional Genetic Counseling Programs. Although the figures may not be exact, they are close, demonstrating two general principles: 1. Primary physicians are critical to the initiation and direction of the genetic counseling effort, i.e., it is they who are most important for implementing the genetic approach to human disease. 2. Genetic counseling and genetic health care in general are not privy to the geneticist. These should be obvious truths, yet they are often overlooked by health care educators and planners.

TABLE 10-2
General Principles as Guidelines for Genetic Counseling

1. Accurate diagnosis a *sine qua non*
2. Central importance of communication
3. Nondirectiveness of the counseling
4. Respect for and adjustment to the family's psychological and emotional turmoil
5. The family unit as a focus of health care
6. Positive contributions to patient management
7. Genetic counseling not circumscribed, isolated clinical effort
8. Genetic counseling is time consuming
9. A team approach is optimal

ber's time and optimal availability of this time to the families being assisted. The clinical team usually consists of at least one physician geneticist, a genetic associate or nurse-geneticist, and a clinic secretary. A social worker and cytogenetics laboratory personnel may also be part of this team. The value of discussing each case with professionals to reach a consensus diagnosis and obtain feedback on the counselor's interaction with the family cannot be overemphasized.

Preliminary Genetic Counseling

In keeping with the schema of Fig. 10-2 and Table 10-1, preliminary genetic counseling involves the initial contact with the proband and/or consultand, portions of the genetic evaluation, and the earliest (usually tentative) efforts to explain the situation to the patient or family. This early contribution to the genetic counseling process again points up the need to integrate genetics at all levels of health care. If the preliminary counseling is not handled well, the definitive counseling will suffer. As the genetic evaluation proceeds, the counselor can get to know the family members so that the definitive counseling is presented in a way that maximizes its usefulness to them. If family members are angry, for example, it is incumbent upon the counselor to deal with their anger so that it does not become an impediment to using the information that is ultimately provided. During the preliminary counseling, efforts are made to determine whether there are any special emotional or domestic problems (e.g., previous misdiagnoses, other ill family member, parental depression, pending divorce, financial difficulties) that may complicate the counseling process. Family members, especially parents of severely affected or deceased children, must have the opportunity of leisurely discussion of the situation as they see it. In turn, the counselor must be very careful to exercise restraint in suggesting a diagnosis and giving advice before the evaluation is completed.

In regard to these emotional aspects of counseling, two things should be kept in mind: 1. *Nonverbal communication* helps determine receptiveness to explanations later on, so careful attention to all levels of interaction in the preliminary stage will enhance the counselor's ultimate effectiveness (2,3). Clock-watching, for example, undermines any semblance of an unhurried atmosphere and raises a question about the counselor's depth of concern. 2. Virtually every individual or couple presenting for genetic counseling has some degree of underlying *guilt, anger, and/or frustration* which, if present, must be dealt with in the preliminary counseling so as to reduce its adverse influence in later sessions. The point is not to engage in psychodynamics, but to be prepared for such reactions and discuss them openly. If these elements are dealt with directly, then counseling can more effectively deal with the genetic details. Many couples, for example, direct anger at the counselor; ordinarily this should not be taken personally. Many mothers, and often fathers as well, reveal a sense of guilt by repeatedly trying to attribute their child's problem to something they did during or before the pregnancy. To be distressed and frustrated over having a severely handicapped or stillborn child is certainly realistic, and these emotions should be apparent. If they are not, it may be helpful to probe for them.

Different members of the genetic counseling team usually take part in the preliminary counseling effort, especially the clinic secretary, the counselor and, the genetic associate. Genetic associates and nurse-geneticists have an important function in these early stages: They participate in the evaluation (history-taking, pedigree construction, etc.); they are in a particularly good position to get to know the family, allowing family members to ask questions about which they might otherwise be reticent (4); and they allow the team as a whole to counsel more families in a given time period. (See also Chapter 16.)

In summary, the preliminary counseling is characterized by a team approach, overlap with the evaluation procedures, restraint from giving advice prematurely, efforts to allow family members to express themselves, and the counselor's getting to know the family in order to maximize the effectiveness of the final counseling.

Definitive Genetic Counseling

The primary goal of genetic counseling is to provide information that could be useful in making rational decisions about childbearing. Secondary, though no less important and often more immediate, goals are emotional and psychological support, improved management of rare disorders, and proper referral for implementation of families' decisions (Chapter 11). To be effective in achieving these goals, the techniques of genetic counseling must take into account the fundamental role of communication and the psychological elements of the situation. Some authors even refer to genetic counseling as *psychologic medicine* (5). However, genetic counseling is sought for many different reasons and, from one situation to another, both the specific goals and the techniques used to attain them will vary.

Once the final disposition has been formulated, emphasis shifts to conveying information to the affected individual and/or family members. The definitive counseling provided by genetic counselors is influenced by all nine principles enumerated earlier, but special attention is given to two sets of considerations: 1. the content, form, and manner of communication, which must be nondirective; and 2. the psychological and emotional components of the situation. The information must be provided in a nonhurried and sincere manner, with a minimum of technical jargon. When technical terms such as autosomal dominant, mutation, and carrier are used, they must be adequately defined. Questions by family members should be encouraged and answered as fully as possible. If the counselor hasn't had to answer "I don't know" several times, the family probably has not been asking enough questions. Deliberate efforts to deal with the family's presumed emotional turmoil must be made at this stage. If it is true (and almost always it is), it is worth saying several times to a couple that "there was nothing you did or did not do that has anything to do with your child having this condition." Making it easy for family members to cry may help them come to grips with the situation and also reveal any excessive or unusual emotional components that require referral for further assistance. When the essential information has been imparted, family members should be asked to repeat back the major points, so that it is at least clear that they heard correctly. They should be encouraged to ask additional questions at the time or subsequently, in person, over the phone, or by mail. A copy of the genetic counseling report or a separate summary letter is sent to the family shortly after the counseling session.

Although the counselor cannot avoid influencing a family's decisions, he must let family members know that the choice between alternative courses of action is theirs to make. He ensures that they have all the needed information available but does not tell them how to use it. The pressure on the counselor to make such decisions may be subtle or obvious, as in the query, "What would you do if you were me?" However expressed, this question has to be turned back to the family. Ordinarily, at least one followup visit is warranted—to promote or consolidate the family's understanding and decisionmaking, to give emotional support, to help implement decisions, and to assist in the affected patient's management.

Clinical Examples

Several real counseling reports with falsified names and places are provided below to exemplify how the foregoing considerations are incorporated into the counseling process. Note that in each instance a copy of the report was given to the family.

CASE NO. 1

DIAGNOSIS: PREVIOUS CHILD WITH TRISOMY 18
GENETICS: CHROMOSOMAL; 47,XY,+18

Mr. L. and Mrs. D. Bloss are here today out of concern for the heritability of the disorder suffered by their son, Leonard (date of birth: September 2, 1974). They were referred in association with their physician, Dr. D. Holf.

Leonard was born of an unremarkable pregnancy that was diagnosed by a urine test. Delivery was one month premature and in the newborn nursery a weak cry was noted and oxygen was administered. Problems were recognized early on and he expired on September 4, 1974. Prior to his demise a sample of blood was obtained for chromosome analysis and our laboratory demonstrated the presence of an extra chromosome #18.

Leonard was born when his mother was somewhat less than 37 years old. The *family history* is unremarkable for other individuals with chromosomal anomalies.

A copy of the chromosome analysis (karyotype) was shown to Mr. and Mrs. Bloss. The association between this type of chromosome abnormality (a change in chromosome number) and her age was clarified. It was pointed out that for a woman between the ages of 35 and 40 years, there was approximately a 1% (1 out of 100) risk of this type of disorder occurring. We then went on to indicate that she would have approximately a 1% or so recurrence risk for the development of a similar type of abnormality in subsequent pregnancies. She understood that the most usual types of abnormalities were an extra chromosome #21, #13, or #18. We explained that there was probably increased risk for the offspring to their other two sons, although the specific magnitude could not be clarified at this time. That could certainly be discussed in the future when those two boys are ready for marriage.

We spent most of our time today discussing various aspects of amniocentesis for prenatal diagnosis of chromosome disorders. It was pointed out that in the context of the 1% risk for a similar type of chromosome abnormality, there was a risk perhaps as high as 1% for an unexpected miscarriage in association with the amniocentesis, although we were generally encouraging and enthusiastic about this procedure. They would very much like to have a third healthy child, but I think without amniocentesis they would be reluctant to make another effort. Thus, I indicated that we would do all we could to avail them of this diagnostic procedure and assist them with implementing any decisions made as a result of the tests done on the specimens.

We discussed that there was nothing from a genetic standpoint to suggest a delay in pregnancy if indeed that was what they wished from their own emotional, social, and domestic standpoint. In particular, we considered that, taking into account the social aspects of their ages, as distinct from medical or obstetrical implications thereof, if they were to have more children, the sooner might be the better. In response to multiple questions, it was made clear that the amniocentesis would provide results and indications of a normal fetus in terms of the specific tests that were to be performed. Thus, a normal chromosome analysis on the fetal cells would indicate a chromosomally normal fetus. Although this would in no way rule out the possibility of any other kind of problem, there was no increased expectation over the general population. I believe that all these points were well understood. We also spent a fair amount of time discussing some of the details about the procedure, including the need for ultrasound to localize the placenta and to rule out twins (there is a positive twin history), and the means by which the needle is placed into the amniotic cavity through the abdomen under local anesthesia, fluid withdrawn, and the cells and fluid itself examined in a variety of ways to assess the fetus.

In the event that Mr. and Mrs. Bloss choose to have another pregnancy, it would be appropriate for us to get together again to talk over further the details regarding amniocentesis. I believe that they now know the general indications as well as the specific indication in their own regard, the timing for the procedure (approximately 15 weeks gestation), what is to be gained from it, as well as the relatively minor risk.

A copy of this dictation will be made available to Dr. Holf as well as to Mr. and Mrs. Bloss. A copy of the chromosome analysis will also be made available to Mr. and Mrs. Bloss.

CASE NO. 2

DIAGNOSIS: LEPRECHAUNISM (DONOHUE SYNDROME)
GENETICS: AUTOSOMAL RECESSIVE

Mr. T. and Mrs. K. Welt are here today with their daughter, Ann. They have been referred to us by Dr. P. Stevens. They are concerned about the prognosis and heritability of the disorder which afflicts Ann.

Ann was born of an unremarkable pregnancy. There were no

significant teratogen exposures. Delivery was unremarkable; birth weight was 9 lb. 15 oz., and birth length was 21″. Because of cyanosis and respiratory difficulties at the time of birth, resuscitation was necessary on several occasions. She remained in the hospital for the first five weeks of life. Immediately from the point of discharge to home she was noted to be an irritable child with difficulty feeding. There has been constant lag in her developmental milestones. She is only now beginning to sit unassisted (26 months). She does reach for objects, although feebly. Her vision and hearing ability are thought to be intact.

Evaluations on several previous occasions had not substantiated a specific diagnosis. However, Dr. Stevens has felt all along that Ann's problem is properly designated by the term Leprechaunism, or Donohue syndrome. The main purpose of this visit today is to corroborate that diagnosis insofar as possible.

The family history is unremarkable for similarly affected individuals. However, Mr. Welt's mother's maiden name was Prefak, and Mrs. Welt's mother's maiden name was Prefak also. No specific blood relation between those two individuals is known, but they are from the same ethnic/national background. It is of interest as well that Mr. Welt's parents are themselves blood relatives, reportedly being related as "third cousins." Ann had a brother born in October 1970 who died at two months of age with a diagnosis of "crib death." A sister, Roberta, was born October 1, 1975.

On *physical examination*, Ann had the appearance of a child with severe failure to thrive. The head circumference was 41 cm (much less than the third percentile) and though small, was relatively larger than the rest of the body would lead you to expect. Her length was 63 cm, which again is much less than the third percentile. The cranium itself was not remarkable except for a relative prominence of the occiput. There was no asymmetry. The facial appearance was striking, with an unexpectedly aged appearance. There was an increased amount of fine, blond hair over the face. The eyes were wide open with the sclerae prominent. The inner canthal distance was 28 mm, the outer canthal distance 70 mm, and the inter-pupillary distance 35 mm. These measurements are indicative of hypertelorism. The palpebral fissures each measured 21 mm. The tip of the nose was relatively prominent. The mouth was relatively large for the size of the face. The tongue was unremarkable. The external ears were relatively large for the overall size of the head. The helices were overfolded. There were indentations on the outer segment of the lobule and there was a series of small pits on the posterior aspects of the lobules, more prominent and numerous on the right. Externally the eyes were not otherwise abnormal—there were no corneal opacities, the irises were unremarkable, and the pupils were round, central, and reactive to light. The neck and remainder of the spinal column were unremarkable except for a lumber kyphosis due to muscular hypotonia. The chest was small, but the sternum was proportionate in length to the rest of the body. The breasts were not remarkable. The heart and lungs were clear to auscultation. The abdomen was free of masses or organomegaly, although the liver edge could be palpated at the right costal margin. The skin had an unusual yellowish tinge which was similar to but not equivalent to icterus. The genitalia were unremarkable. The anal orifice was anteriorally displaced. The upper extremities were unremarkable in terms of morphology. It was very difficult to read her dermatoglyphic pattern. There were no simian creases although there was a Sydney line on the right, associated with a distal axial triradius; on the left there was neither a simian crease nor a Sydney line, and there was also a distal axial triradius. The lower extremities were unremarkable except for the marked wasting as noted for the upper extremities. There was overall a marked decrease in subcutaneous tissue. This made the skin appear as though it were redundant over all of her body. There was prominent venous distension in the distal extremities. She was profoundly hypotonic. She apparently could see and hear. She was an inattentive, listless child.

She was seen by Dr. J. H. and other members of the Clinical Genetics Center, University of Wisconsin—Madison, who were attending in the clinic with me today. Our concensus was that the diagnosis of Leprechaunism applies on her behalf. We felt that this was the case in terms of her clinical course and physiognomy, particularly her facial appearance, the ear pits, and her loss of subcutaneous tissue. It is somewhat unusual, however, that her birth weight was as high as it was. Most individuals with this disorder have a prenatal onset, with a low birth weight. Our initial discussions with Mr. and Mrs. Welt had three particular components. First, we indicated an agreement with Dr. Steven's diagnosis of Leprechaunism. This in turn was indicated to have at least two consequences. One of these is the poor long-term prognosis, particularly for mental capacity. I think the Welts are coming to grips with this problem already. We indicated that there was virtually nothing that could be done to afford any improvements. I believe that this was reasonably well understood. The other consequence was the possibility of recurrence. This respects the possible consanguinity of Ann's parents and evidence from the literature to indicate that this is an autosomal recessive disorder. Thus, there could be a 1-in-4 recurrence risk.

All in all, I think the situation was well understood. Their main concern now was whether or not anything could indeed be done on behalf of Ann. We indicated that there was not any specific thing available under ordinary circumstances and, further, if anything were done it would almost necessarily have to be on a research or investigative basis. Under those circumstances, some preliminary arrangements would be required to make such available. I indicated we would be in touch with Dr. Stevens and assess his feelings in this regard. If both he and the Welts felt it would be worth pursuing, then perhaps some arrangements could be made for proper investigation, as for example at the National Institute of Health.

A copy of this dictation will be made available to Dr. Stevens and to Mr. and Mrs. Welt.

CASE NO. 3

DIAGNOSIS: THREE CONSECUTIVE MISCARRIAGES; MOST RECENT ABORTUS TRIPLOID.

GENETICS: CHROMOSOMAL; RECURRENCE RISKS NOT CLEAR.

Mr. E. and Mrs. M. Jones are here today for genetic counseling regarding three consecutive miscarriages. On December 12, 1975 Mrs. Jones miscarried after approximately one week of bleeding per vaginum. On December 6, 1975 she had received an injection ostensively to assist her through pregnancy in view of the previous history of miscarriages. Her LMP had been October 23, 1975. The intact fetus was brought to us in the chromosome laboratory; we washed it with sterile antibiotic-containing medium, and explanted it into fibroblast culture. The cells grew well and we were able to document a nonmosaic triploid chromosome constitution for the fetus. These 69 chromosomes were associated with a sex-chromosome constitution of two X's and one Y.

The Jones have a normal, healthy son, born October 12, 1972. In March 1974 there was a spontaneous abortion at about 3½ months gestation. In April 1975 there was a second spontaneous abortion estimated to be at about seven weeks. The abortion of December 1975 was the third abortion for this couple and their fourth pregnancy. By history, there was no teratogenic or other environmental exposure including drugs, x-rays, and infections. Chromosome analyses were not performed on the previous two abortions. Both Mr. and Mrs. Jones indicate that their health is otherwise generally good and they have no specific complaints. They both have normal karyotypes. The family history is unremarkable for other individuals having multiple miscarriages.

We discussed at length the triploid nature of the aborted fetus. We went on to say that we were unsure how this related to the other two miscarriages, although there is precedence for recurrence of triploidy among consecutive multiple miscarriages to a couple. There is thus some reason to expect that this type of problem could happen in subsequent pregnancies. In addition, a number of triploid fetuses have reached term to die in early infancy.

An immunologic evaluation for usual or unusual antibody incompatibilities was unrevealing. Thus, from the serologic approach, there was no apparent basis for their multiple miscarriages. In particular, there was no evidence of Mrs. Jones having anti-T_{Ja} antibody.

While the risk for recurrence is probably low, we ordinarily make available to couples such as Mr. and Mrs. Jones amniocentesis for prenatal diagnosis of chromosome disorders in any subsequent pregnancy. It was emphasized that this availability was based on the general principle that having had one child or fetus with a chromosome abnormality puts them at increased risk. Under those circumstances, amniocentesis is a reasonable alternative. I emphasized that I was not stressing that they utilize this kind of approach, but rather that they recognize its availability to them. We discussed only briefly the various aspects of prenatal diagnosis. We went over the main concern in terms of risks, mainly the possible association with unexpected miscarriages. They understood that it is not clear at this time that there is any specific adverse consequence to the procedure.

We left off that they would be in touch with us in the event of a subsequent pregnancy if they were interested in availing themselves of amniocentesis for prenatal diagnosis. At that time, we will assist them as they see fit. In the meantime, a copy of this dictation will be made available to them and to Dr. Gribb. Each will also get a copy of the chromosome analyses.

CASE NO. 4

DIAGNOSIS: WILLIAM WITH DUCHENNE MUSCULAR DYSTROPHY. CONCERN FOR CARRIER STATUS IN HIS SISTER, BARBARA.

GENETICS: X-LINKED RECESSIVE; APPROXIMATELY 6% CHANCE THAT BARBARA IS A CARRIER.

Mr. T. Code and his daughter, Barbara, are here today for a followup visit, accompanied by Ms. Lenz, psychiatric social worker, to discuss the conclusions regarding the likelihood that Barbara is a carrier for Duchenne muscular dystrophy.

At our previous visit of March 3, 1976 a detailed family history was obtained, the data from which indicate that almost certainly Mrs. Code did not inherit the mutation from her mother. This, in turn, means that we are concerned about the origin of the Duchenne muscular dystrophy in either Bill himself, or in his mother. Using limited information having only to do with genetic principles, the conclusions would be that Barbara has a 1-in-3 chance of being a carrier. That estimate is much too high. Utilizing additional data, the risk is actually 6% for her being a carrier and 3% that she have an affected son.

Mrs. Code has three normal boys in addition to her one affected son. According to tables available to make these determinations she thereby has a .58 likelihood of being a carrier. She has had three normal serum creatinine phosphokinase (CPK) determinations over 2½ weeks time. Barbara's determinations on the same days were within normal limits. If we consider a 60% reliability of the serum CPK serial determinations, this in turn means that there is the possibility of a 40% false-negative (i.e., that the person could still be a carrier and show up as having a normal serum CPK level). If we then take the .58 likelihood of being a carrier and multiply times .40, then the combined likelihood that Mrs. Code is a carrier is .23. The likelihood of her passing the mutant gene to Barbara is .5, and .5 × .23 is equal to .16, which is the likelihood that Barbara is a carrier, based only on data relevant to her mother. If we now consider that she has had three normal CPK determinations, and multiply by the same .40 false-negative coefficient, then the likelihood that Barbara is a carrier using all the data is .06. In turn, the likelihood for having an affected male child is half of that, or .03. Thus, there is approximately a 3% risk for Barbara's having an affected son (or a 1.5% risk for having an affected child).

We discussed at length that a 3% recurrence risk for Barbara's sons was helpful in that she could realize that there was no particularly high risk. On the other hand, when it came time to think specifically about having a child, then this 3% risk might have a somewhat different meaning. In this context, we could then consider the availability of prenatal diagnosis utilizing amniocentesis for sex determination. If the fetus was a male, then termination of pregnancy could be afforded, based on the 3% risk likelihood. This is, of course, a drastic consideration since that would mean that there would have been a 97% chance of terminating a pregnancy for a normal male fetus. So, rather than pursue this particular avenue, I suggested that recounseling be afforded when Barbara marries and is considering procreation. At that time, an improved diagnostic approach might be available, allowing for specific prenatal diagnosis. I believe that she and her father understood all of these points quite well.

We spent considerable time dealing with the emotional impact that various aspects of Duchenne muscular dystrophy have been having on the family.

A copy of this dictation will be made available directly to Barbara and another copy to her parents. We discussed that it would be appropriate for her parents to share this information with their sons. In particular it is important for them to know that they are at risk neither for passing on the carrier status nor the disorder. Again, I believe all of this was quite well understood.

CASE NO. 5

DIAGNOSIS: PREVIOUS CHILD, EXPIRED WITH MICROCEPHALY, FRONTAL AND OCCIPITAL ENCEPHALOCELES, SACRAL MYELOMENINGOCELE, FOREARM ABNORMALITIES, ABNORMAL DERMATOGLYPHICS, MICROPHTHALMIA, AND BILATERAL SEVERE CLEFT LIP AND PALATE.
GENETICS: UNKNOWN; VERY POSSIBLY TRISOMY 13 (PATAU SYNDROME). PARENTAL CHROMOSOMES NORMAL.

Mr. F. and Mrs. T. Sepp are here today to discuss the possible diagnosis and heritability of the disorder noted above. They came to us after my discussion with the physician who delivered the baby, Dr. L. Roberts. I first saw their infant on Monday, October 9, 1975 when I examined her postmortem. The essential findings are listed in the diagnosis above and the complete physical examination has been detailed in my dictation of that same day. As a consequence of my contact with Dr. Roberts, peripheral blood specimens had been drawn from both Mr. and Mrs. Sepp for a chromosome analysis on the basis of the remote possibility that either might be a balanced translocation carrier involving chromosome #13.

The pregnancy was generally unremarkable except for the fact that it was diagnosed by an endocrine provocative test. There were no other significant teratogenic exposures, although Mrs. Sepp indicates that she had a "cold" throughout much of the early pregnancy. She was treated with penicillin for it. Delivery was at term and relatively unremarkable. Birth weight was 7 lbs. 1 oz. Death was shortly after birth.

The family history is essentially unremarkable except for one uncle to Mr. Sepp who is said to have had an "open spine." No further details are known in that regard. There are no known instances of chromosome abnormalities or multiple congenital malformations obvious at birth. Mr. Sepp had had a vasectomy following the conception of this affected child. They were, nonetheless, quite interested in pursuing the genetic counseling, because of the various possibilities of subsequent pregnancies, including by such means as surgical correction of the vasectomy, artificial insemination by donor, remarriage, etc. We also considered that if indeed either of them was a carrier of a chromosome rearrangement, it would be important to each of their siblings and other relatives. Finally, both were interested to know, in any case, whether or not there was anything they had done which might have contributed to the baby's condition.

First, I indicated that although chromosomes were not done, the physical findings on the baby were so indicative of trisomy 13 or Patau syndrome that it was reasonable to proceed as though that were the diagnosis. Given that, then the next thing was to determine whether or not it was due to literal trisomy 13 or whether there might be an unbalanced translocation involving chromosome #13. Since chromosomes were not available on the baby, the next thing was to examine the chromosomes of both parents. I explained to them that I had seen cases where indeed one of the parents had turned out to be a balanced carrier for a child presenting in much the same way as their daughter. The chromosome analyses done on both Mr. and Mrs. Sepp were completely within normal limits, i.e., there was no evidence of any kind of chromosomal rearrangement, including specifically any involving the #13 chromosome. Based on these data, then, there was no reason for Mrs. Sepp's sister, who is currently pregnant, to be particularly concerned. Likewise for their cousin who is to deliver in February 1976.

We then went on to consider that if we continued to presume that this was trisomy 13, then in the event of any subsequent pregnancies, it would be reasonable to perform amniocentesis for the prenatal diagnosis of chromosomal disorders and neural tube defects. This would involve analysis of the fetal cells for the chromosome abnormality and analysis of the amniotic fluid for alpha-fetoprotein for neural tube defects. We did not go into all the details about amniocentesis, although I did indicate there was no question but that this procedure was appropriate for Mr. and Mrs. Sepp in the event of a pregnancy.

I was most explicit and emphatic that there was nothing they did or did not do which in any way contributed to the problems that were seen in their deceased baby. Likewise, I indicated that there was no way to specifically implicate the purported "viral infection" early in the pregnancy. We discussed further that although provocative pregnancy tests are anachronistic and that there are, at times, reasons to be concerned that those tests can cause problems, this was not the case here. I believe they were relieved to have acknowledged that nothing they did caused this problem.

All in all, I think they understood the situation quite well. It would be most appropriate in the event of any subsequent pregnancies that we see Mr. and Mrs. Sepp again to go over the details of amniocentesis, whether or not they elect to utilize that avenue. The point would be to make sure that they knew the details so that their decision could be fully informed. A copy of this dictation will be made available to Mr. and Mrs. Sepp as well as to Dr. Roberts.

CASE NO. 6

DIAGNOSIS: CONRADI DISEASE

GENETICS: AUTOSOMAL DOMINANT; NEW MUTATION?

Mr. A. and Mrs. B. Loft are here today with Sally out of concern for the heritability of Conradi disease, which afflicts Sally. I originally saw Sally and her parents the evening of December 4, 1975, shortly after Sally's birth at County Hospital. At that time, I corroborated the clinical diagnosis of Conradi disease and discussed the range of possible clinical stigmata which were indicated to be variable, depending on the specific disorder and mechanism of inheritance. At the time that I spoke with the Lofts there was a considerable amount of emotional turmoil.

Since then, I have shared the clinical story and x-rays with Dr. J.O., Department of Medical Genetics, University of Wisconsin—Madison. We agreed that, based on the clinical and radiographic data, this is the autosomal dominant variety of Conradi disease. Further, we discussed the extreme variability of this mutation from one individual to another. Thus, it is not entirely out of the question that either Mr. or Mrs. Loft might have the mutant gene and not be expressing it very much, if at all. The importance of that consideration is that we cannot rule out the unlikely possibility of recurrence of this disorder among future siblings to Sally.

This point in particular was made to Mr. and Mrs. Loft and I think they understood it reasonably well. We approached it from several vantage points: the main emphasis was that there is no basis for saying that she will literally grow out of the disorder, but on the other hand this is not a progressive disorder and it is relatively benign. In particular, there is no associated mental retardation. Mr. and Mrs. Loft were glad to hear these tidings. However, they did have a number of specific questions about Sally: they were concerned about the limitation of motion at the hips (extension, external rotation, and abduction) and the apparent decreased use of her right upper extremity (reaching for or warding off things with the right side).

On physical examination Sally looked to be a happy, vigorous child. The skin was unremarkable, in contrast to the erythematous hyperkeratotic rash that was present at birth. There was indeed modest limitation at the hips with the movements as noted above. There was no doubt that she used the right upper extremity considerably less than the left for reaching and holding onto things. There was also a difference in the posturing of the fingers, with less inter-phalangial flexion on the right as compared to the left. There was a moderate amount of ulnar deviation on the right as compared to the left.

In general, I was encouraging to Mr. and Mrs. Loft about the long-term outlook for Sally. I also indicated in review that it was not possible to say that this disorder wouldn't happen again in future offspring to them, although the risk was relatively small, most probably well below 10%. I emphasized that the main reason for hedging had to do with the incomplete expression of this disorder on some occasions. A family history indicated that there were no other family members who could even remotely be considered to be affected. This includes a substantial number of aunts, uncles, and cousins on both sides.

We then went on to consider that when it is time for Sally to have children of her own, she would have a 50% risk for passing on the mutation and, in turn, a proportion of them would be at risk for developing the fullblown syndrome as she did.

In summary, I believe that all of this was reasonably well understood. However, I think that Mr. and Mrs. Loft are still experiencing a fair amount of turmoil and they could use support and close contact along the way. It would appear that they are getting this in their comfortable association with Dr. Williams.

A copy of this dictation will be made available to Dr. Williams as well as to Mr. and Mrs. Loft and to Sally's County Hospital chart.

CASE NO. 7

DIAGNOSIS: SHORT STATURE, DISTINCTIVE FACIAL APPEARANCE, SMALL GENITALIA, ABNORMAL DERMATOGLYPHICS; POSSIBLY THE NOONAN SYNDROME.

GENETICS: UNCLEAR AT THIS TIME; AUTOSOMAL DOMINANT, NEW MUTATION?

Mr. C. and Mrs. J. Plenum are here today for a followup visit with Jeff.

The first thing pointed out was the fact that the buccal smear and chromosome analysis on Jeff's peripheral blood specimen were normal. Thus, the chromosome abnormality which I thought was a good possibility did not turn up. I indicated that under these circumstances I was unwilling to make any specific diagnosis, although I did indeed strongly entertain the possibility of the Noonan syndrome. My reluctance for focusing on that particular diagnosis had to do with the absence of mental retardation, relatively severe ptosis, or hearing deficit. However, we went on to indicate that perhaps by analogy some of the same considerations as for the Noonan syndrome would have to apply. The two most important ones would be the anticipated lack of complete sexual maturity, perhaps associated with sterility, as well as the possibility of an autosomal dominant mode of transmission. This latter would entail a 1-in-2 recurrence risk for each offspring to Jeff, in the event of procreation. I believe that the reason for not making a specific diagnosis was clear, as was the need to make some tentative considerations for realistic expectations in the future. We went on to discuss that if indeed hypogonadism or incomplete sexual maturity was to be a part of Jeff's life, then efforts to prepare him for that were in order; for example, minimizing discussion about expectations for grandchildren, marriage, etc. We also discussed that in the event that Jeff did not show any clear signs of entering puberty by age 13 (in the context of his father having begun puberty around age 12 years), then it would be reasonable to do an endocrinologic evaluation to determine whether hypogonadism was to be involved and also whether hormonal therapy would be appropriate. All in all, I think Mr. and Mrs. Plenum understood the situation quite well.

We spent a fair amount of time discussing the overall maturity of Jeff and how well he has handled his distinctiveness.

We also went over the original reason they were concerned about the need for medical attention, that is, his school's concern for his scoliosis. Mr. and Mrs. Plenum and I discussed that although a minor finding, in the context of his situation it made a somewhat greater impact.

I indicated to the Plenums that I would like to see Jeff for a followup visit in about one year's time, and they were glad of that. A copy of this dictation will be made available to them as well as to Dr. Hayes.

Effectiveness of Genetic Counseling

Efforts to answer questions that are consistently, intelligently, and appropriately asked by patients in a clinical context need no justification. That is a major task of clinicians: to deal with the health concerns of their patients. This includes the question, "Will it happen again?," which underlies so much of clinical genetics and relates to almost all disorders. As a formalized way of answering inquiries about occurrence and recurrence risks, genetic counseling is an effective mode of health care: it provides reasonable answers to reasonable questions. Yet with increasing frequency the "effectiveness of genetic counseling" is being called into question.

I believe this skepticism is based on several misconceptions: 1. It treats genetic counseling as an end product, not a process. 2. It fails to distinguish between the process in a specific situation and in a large well-organized genetics program where the efficacy of the program (rather than individual situations) may be the real question. 3. It is actually concerned with the specific techniques of the process and not the process itself. That is, the process is effective if it answers the questions asked. However, do the techniques involved change the process? This should be the central concern about counseling effectiveness. 4. It attempts to judge genetic counseling's effectiveness in terms of how a third party feels the information provided should be used, i.e., "Did all the families 'comply with' the counseling?"

Conclusion

The genetic approach to human disease is an attitude, an inclination to respond in a certain way. Genetic counseling is one aspect of that response. By acquiring an ability with genetic counseling techniques, clinical practitioners can make the genetic approach more available to their patients. However, not only should practitioners appreciate this opportunity; health care educators and health care planners should also be made aware of such techniques. Clinical genetics training programs in all health care curricula are necessary to ensure the maximal impact of modern genetics both for on-going management and primary prevention. Regional centers for genetic counseling, funded through state and federal agencies with local community support, are also necessary to assure the uniform, widespread availability of these services.

Cited References

1. Committee on Genetic Counseling, Am Soc Hum Genet: Genetic counseling. *Am J Hum Genet* 27:240, 1975.

2. Krumboltz J, Varenhorst B, Thorensen C: Nonverbal factors in the effectiveness of models in counseling. *J Counseling Psychol* 14:412, 1967.

3. Hinde RA (ed): *Nonverbal Communication.* Cambridge, Cambridge University Press, 1972.

4. Kelly PT: The role of the intake interview in genetic counseling. *Am J Hum Genet* 27:53A, 1975.

5. Hecht F, Holmes LB: What we don't know about genetic counseling. *New Engl J Med* 287:464, 1972.

General References

A. Fotheringham JB: Counseling with families of children with learning disorders. *Canad Fam Phys* 21:82, 1975.

B. Fox AM: Families with handicapped children—a challenge to the caring professionals. *Community Health* 6:217, 1975.

C. Fraser FC: Genetic counseling. *Am J Hum Genet* 26:636, 1974.

D. Kushnick T: When to refer to a geneticist. *JAMA* 235:623, 1976.

E. Milunsky A: Genetic counseling: Principles and practice. in Milunsky A (ed): *The Prevention of Genetic Disease and Mental Retardation*, Philadelphia, WB Saunders, 1975, pp 64-89.

F. Murphy EA, Chase GA: *Principles of Genetic Counseling*. Chicago, Yearbook Medical Publishers, 1975.

G. Murphy EA: The rationale of genetic counseling. *J Pediatr* 72:121, 1968.

H. Pueschel S, Murphy A: Counseling parents of infants with Down's syndrome. *Postgrad Med* 58:90, 1975.

CHAPTER ELEVEN

THE ROLE OF GENETIC COUNSELING IN FAMILY AND PATIENT MANAGEMENT

Introduction

While the general function of genetic counseling is assessment and communication of risks for future offspring, there are two additional goals: 1. clinical management of the affected patient(s); and 2. clinical management of family members (e.g., emotional support). The relationship of these two goals to the primary function will vary considerably from one family to another. For some families concern about procreation may be moot, and current management of the affected individual is the most important consideration. Others may already be knowledgeable about the genetics and clinical course, but seek counseling for psychological support, or administrative assistance in seeking to adopt a child. That is, the primary purpose of genetic counseling in general is not always the primary goal of the consulting family. Actually, the family members' primary goal is not always known to themselves, although as a starting point it is reasonable to assume that the family's major concern is similar to the counselor's. Knowing when this assumption is wrong can be important. Recognition of the family's genetic concerns, whatever they are, may be the first step to implementing the genetic approach to human disease. Whoever recognizes the concern and does something about it, directly participates in the genetic counseling process.

The purpose of this chapter is to describe how the counseling process contributes to family and patient management. In particular we shall consider the role of support, alternatives to procreation, clarification of the disorder's natural history, and therapy.

Primary, Secondary, and Tertiary Prevention

A recurrent theme in this volume has been the preventive potential of the genetic approach to human disease. Prevention at all three levels defined in Table 11-1 may result from implementing this approach, though the most desirable types are clearly primary and secondary. Primary prevention will be discussed in the next section and therapy of genetic disorders in Chapter 14. Tertiary prevention will be considered below in terms of support for family members; otherwise it is mentioned here mainly to indicate that it is the only level of assistance possible unless there is a specific genetic diagnosis.

For the vast majority of genetic disorders, primary prevention can be implemented only after an affected child has been born and the disorder is obvious. There are two important exceptions to this, however: prenatal diagnosis of chromosome disorders associated with maternal age of 35 years or more, and carrier detection by means of widespread screening programs (e.g., Tay-Sachs disease, sickle-cell anemia and alpha-1-antitrypsin deficiency). While carrier detection programs for sickle-cell anemia and alpha-1-antitrypsin deficiency may also pick up affected individuals who have mild symptoms, other screening programs are capable of detecting only affected patients—e.g., phenylketonuria (PKU), galactosemia. All things considered, primary prevention currently has its biggest impact through age-indicated prenatal diagnosis and recognition of the genetic basis of disorders afflicting patients already born. In both instances it is the primary-care physician who assumes the pivotal role in seeing to it that families have access to the preventive methods (Table 11-1).

Genetic Alternatives

Given the possibility for occurrence or recurrence of a disorder with a significant burden, the couple involved may be able to avoid or reduce the risk. This can be done by contraception (including abstinence), adoption, artificial insemination by donor (AID), or prenatal diagnosis with the option to terminate the pregnancy. Each of these tactics, however, has limited applica-

TABLE 11-1
Preventive Levels for Genetic and Congenital Disorders

Primary prevention. Preventing or avoiding the problem in the first place
1. Immunization against infectious teratogens (e.g., Rubella virus).
2. Avoidance of other environmental agents (Chapter 7).
3. Prevention of pregnancy.
4. Artificial insemination by donor.
5. Prenatal diagnosis and termination of pregnancy if abnormality is found.

Secondary prevention. Treatment of the basic problem before additional consequences develop (e.g., dietary treatment of phenylketonuria (PKU) or galactosemia).

Tertiary prevention. Efforts to minimize the impact of the disorder on the patient, family, and society (i.e., optimal training for a child already mentally retarded from PKU, helping the family to accept and adjust to a malformed infant).

TABLE 11-2

Relative Effectiveness of Alternatives for Avoiding or Minimizing the Risk for Recurrences of a Genetic Condition

	Baseline Risk, %	*Risk with Contraception*	*Risk with AID*	*Risk with Prenatal Diagnosis*
hromosomal				
Aneuploidy, etc.[1]	1–2	0	N	0[9]
Parental carrier,[2] Father	5–10	0	G[7]	0[9]
Parental carrier,[2] Mother	10–20	0	N	0[9]
ingle-gene				
utosomal dominant				
Mother affected	50	0	50%	0[10]
Father affected	50	0	μ[7]	0[10]
utosomal recessive[3]	25	0	μ–1%[8]	0[10]
-linked recessive, affected males[4]				
Mother carrier	50	0	50%	0[11]
Father affected	μ	0	μ	N
-linked recessive, carrier females[5]				
Mother carrier	50	0	50%	0[11]
Father affected	100	0	μ[7]	0[11]
olygenic				
Neural tube defects	5[6]	0	N	0.5%[12]
Other disorders	5	0	?[8]	?[13]

ote that contraception decreases the risks for all classes of genetic disorders, that AID decreases the risks for romosomal and single-gene disorders where the father's contribution is critical, and that prenatal diagnosis decreases e risks for certain instances in all genetic classes. AID = artifical insemination by donor. G = General population risk, specting spontaneous mutations, background carrier frequencies, etc. N = not applicable. μ = mutation rate.

ncludes all spontaneous chromosome abnormalities.

ncludes translocations, certain inversions, and mosaicism.

oth parents carriers.

ikelihood of having an affected male.

ikelihood of having a carrier female.

Aay be higher or lower depending on family history and relationship of consultand to proband.

isk approaches mutation rate if donor has been screened.

espects carrier rate as high as 1/20 for some disorders in general population.

resumes no laboratory error.

Presumes condition diagnosable in utero.

Presumes condition diagnosable in utero, or termination based on sex.

Presumes specific lesion diagnosable in utero; respects 10% false-negative amniotic alpha-fetoprotein determinations.

Efficacy of methods not known.

enatal diagnosis risks noted assume the termination of pregnancy with a positive diagnosis.

tion. Contraception as a genetic measure essentially means no procreation, and this is a drastic step that few couples elect to take unless they can obtain a child by other means, such as adoption. But children available for adoption are relatively scarce. AID has limited social acceptance and is useful only for rare disorders and when the mother's genetic makeup is not the sole or major determinant of recurrence. Prenatal diagnosis is limited to conditions that are detectable in utero (Chapter 8). Table 11-2 shows the relative effectiveness of contraception, AID, and prenatal diagnosis for changing genetic risks.

AID is underutilized as a method of avoiding or minimizing

recurrence risks where the father's genetic contribution is critical. This includes all instances where the father is a carrier for a chromosome rearrangement, has chromosomal mosaicism, is a carrier for an autosomal recessive mutation, has or is at risk to have an autosomal dominant mutation, or is affected with an X-linked disorder. It is not entirely clear why there is such widespread aversion to this technique as a genetic alternative. The procedures are well established, involve minimal risks themselves, and are easily and inexpensively applied (1,2). Legal sanctions against AID offspring are being or have been removed in many states (2,3). From a genetic standpoint the main additional consideration is to assure that the donor has been properly screened for the condition or carrier status to be avoided. This usually involves a careful history and physical examination, and in some cases laboratory screening (e.g., for Tay-Sachs or sickle-cell carrier status or chromosome rearrangement).

In 1975 we surveyed Denver and other Colorado physicians who provided AID (4). Of the 92 respondents, 79% did or said they would utilize this procedure for genetic reasons and (except for one respondent) would ask a geneticist to screen the donor. Thus, unavailability was not the apparent cause of the procedure's underuse, at least in the survey area. We then began to examine attitudes towards AID among potential and actual users of genetic counseling (4). Our preliminary findings suggest that the main deterrants are based on inaccurate "folklore." Couples are very unlikely to choose AID unless they come to us asking for it or we spend an inordinate amount of time explaining the procedure and dispelling their misunderstandings, which borders on persuasion.

Our experience with AID and prenatal diagnosis has made it clear that geneticists sometimes become advocates of certain approaches. This is something of a dilemma. On the one hand genetic counseling should be nondirective (Chapter 10), but, on the other hand, certain alternatives will not be adopted unless we become advocates. For example, when a 40-year-old woman inquires about prenatal diagnosis, we do not merely say that it is sensible; instead, we make special efforts to provide the procedure. Similar advocacy is necessary for AID to become as useful as prenatal diagnosis. The effect of such directive efforts, however, should be to increase the range of choices rather than to make the choice for the patient or family.

Clarification of the Disorder's Natural History

The ability of clinical geneticists to contribute to patient care by clarifying a disorder's course and prognosis is an increasingly important indication for referral. Predictions about recurrence and clinical course are based on accurate diagnosis. A diagnosis resulting from a genetic evaluation has two types of consequences: 1. It may prompt a search for *associated problems* (i.e., urinary tract malformations with 45,X monosomy, hypothyroidism or psoriasis with the Down syndrome). 2. It clarifies the *expected sequence* of events.

Parents of handicapped infants are most often concerned about whether there will be mental retardation, whether growth will be normal, and whether the child will have a normal life span. Answers to these and similar questions, even if they must be tentative, are very much a part of the genetic counseling process. To go over them in the context of a genetic counseling setting can have a unique benefit: to be advised about genetic disorders by a geneticist is important to many patients and families.

One of the most dramatic ways that a genetic diagnosis can shape management is exemplified by the following situation. A newborn infant with physical features suggesting the Patau syndrome (trisomy 13) is in neonatal intensive care in anticipation of clinical deterioration. There is concern about whether or not certain types of extraordinary surgery should be done, or even whether the child should be in the intensive care unit, considering the large number of other critically ill patients. Under these circumstances, marrow cells are aspirated from the proximal tibia and within 5 hours a chromosomal diagnosis of trisomy 13 is made. The bleak prognosis for this disorder makes it reasonable to avoid extraordinary surgery and to utilize the unit's resources first for other infants facing less certainly tragic outcome. Although the demonstrated chromosomal aberration does not of and by itself make this decision, it is critical to that end. And, if the parents understand the nature of the chromosomal diagnosis, it may well be easier for them to accept the child's death as an inevitable consequence of the disorder.

Therapeutic Considerations

This topic will be dealt with in Chapter 13. It is mentioned here, however, because therapeutic considerations are often an integral part of genetic counseling. The genetic evaluation helps define what is amenable to therapy, and the geneticist has immediate access to resources that might otherwise be unknown or unavailable (an extension of the nidus concept of Chapter 9).

Support of Family Members

Genetic and congenital disorders may provoke intense emotional responses. Most are normal reactions to finding out that a disorder or risk for it exists. Some represent abnormal processes of adjustment to the situation. Among the adverse psychological reactions and adjustments, some reflect domestic problems that make the overall situation overwhelming in very realistic ways. A concerted effort to provide support by the geneticist, primary physician, and social worker is needed in such situations, which occur frequently enough that most major genetic counseling clinics have their own social worker. In other cases the reaction is literally pathologic, becoming a special clinical problem in its own right. Support by the geneticist, primary physician, and social worker plus intervention by a psychiatrist or counterpart is then necessary.

The geneticist often is the person who first indicates or later confirms a diagnosis with grave prognosis. For this reason, if no other, the geneticist must give psychological support to the family as its members adjust to the information he delivers.

Three situations in particular are worth considering because they highlight the geneticist's role in providing support. The first involves making or confirming the diagnosis of Huntington chorea (or a similar neurodegenerative disorder) in a parent of the consultand. This gives the consultand a 1 in 2 risk of also developing the disorder and a 1 in 4 risk at that point of having affected children. The situation can be especially difficult if the consultand already has children. The couple then has to deal with the dread prospect of the consultand developing the disease and perhaps having passed on the mutant gene already. These sessions are often tearful and anguished. The picture is complicated further by the fact that hypochondriacal symptoms may develop and suicide attempts may occur. Regardless of other resources utilized for assistance, our experience has been that at such times the counselor is pivotal in facilitating a realistic acceptance of the situation (5). Referral to local chapters of the Committee to Combat Huntington Disease (6) is also helpful.

A second situation involves the descriptive diagnosis of a congenital syndrome in an infant or small child which leaves the course and prognosis unclear. Frankness, sensitivity, and warmth on the part of the counselor can help the family accept realistic expectations and yet avoid self-fulfilling prophecies about such problems as developmental delay and mental retardation.

The third situation involves prenatal diagnosis by amniocentesis. Currently there is a three- to five-week interim between the time the specimen is obtained and the results are available. This period of waiting to be told whether their fetus shows normal test results is uniquely stressful for the couples involved. It is one of the reasons, as outlined in Chapter 8, that genetic amniocentesis must be part of a comprehensive genetics program to assist and support the prospective parents during this period of turmoil. The need for this kind of assistance to the expectant parents can be so great that most prenatal diagnosis centers employ a special member of the genetic counseling team to be responsible for assisting the couple during this time and for following up with them after delivery. The involvement of the genetic counselor or counseling team with the couple during the waiting period may also be important in other ways, since the prenatal diagnosis process may significantly influence subsequent rearing of the delivered, presumably normal child.

The process of genetic counseling is clearly effective as a formal means of answering questions and providing emotional support. This is true even though questions have been raised as to whether genetic counseling effectively decreases the recurrence of genetic disease. This distinction is important because it explains why genetic counseling may be worthwhile even if further procreation is not intended, and it emphasizes need to measure the efficacy of the counseling separately at each of its levels of function.

Cited References

1. Langer G, Lemberg E, Sharf M: Artificial insemination. *Internatl J Fertil* 14:232, 1969.

2. Beck WW Jr: A critical look at the legal, ethical and technical aspects of artificial insemination. *Fertil Steril* 27:1, 1976.

3. Ciba Foundation Symposium 17 (new series). *Law and Ethics of AID* and *Embryo Transfer*. Elsevier/Excerpta Medica. Amsterdam, North Holland, 1973.

4. Ponce E, Riccardi V: Facilitation interference factors in information processing in genetic counseling. AID as a model. *Am J Hum Genet* 27:72a, 1975.

5. Lynch HT, Harlan WL, Dyhrberg JS: Subjective perspective of a family with Huntington's chorea. *Arch Gen Psychiat* 27:67, 1972.

6. Committee to Combat Huntington's Disease. 250 West 57th Street, New York, NY 10019.

General References

A. Antley MA, Antley RM, Hartlage LC: Effects of genetic counseling on parental self-concepts. *J Psychol* 83:335, 1973.

B. Fraser FC: Genetic Counseling. *Am J Hum Genet* 26:636, 1974.

C. Holloway SM, Smith C: Effects of various medical and social practices on the frequency of genetic disorders. *Am J Hum Genet* 27:614, 1975.

D. Kaback MM: Heterozygote screening for the control of recessive genetic disease. in Milunsky A (ed): *The Prevention of Genetic Disease and Mental Retardation*, Philadelphia, W B Saunders, 1975, pp 90–113.

E. Leonard CO, Chase GA, Childs B: Genetic counseling: A consumer's view. *N Engl J Med* 287:433, 1972.

F. Lynch HT, Krush TP, Krush AJ, et al: Psychodynamics of early hereditary deaths. *Am J Dis Child* 108:605, 1964.

G. Reynolds BD, Puck MH, Robinson A: Genetic counseling: an appraisal. *Clin Genet* 5:177, 1974.

H. Tips R, Smith G, Lynch H, et al: The "whole family" concept in clinical genetics. *Am J Dis Child* 107:67, 1964.

I. Wells RW: Huntington's chorea: Seeing beyond the disease. *Am J Nurs* 72:954, 1972.

CHAPTER TWELVE

ROLES OF NONGENETICISTS AND NONPHYSICIANS IN CLINICAL GENETICS

Introduction

Clinicians at all levels of health care have occasion to consider whether genetic mechanisms may be at play in a given case. This chapter is mainly for clinicians who are not geneticists but who regularly deal with clinical situations involving genetic and congenital disorders. Such persons have diverse backgrounds and are part of varied clinical facilities: dentist's and physician's offices, delivery rooms, newborn nurseries, specialty clinics (e.g., cystic fibrosis, muscular dystrophy), disease-oriented programs (e.g., sickle-cell anemia, hemophilia), adoption agencies, family planning clinics, special education programs, etc. They may take part in genetic counseling in one of three ways: 1. As part of a genetic counseling team. 2. Before and/or after the family is involved with the team (referral, followup). 3. Instead of referral to the team.

Practicing Physicians and Dentists

While dentists and nongeneticist physicians actively and effectively provide genetic health care in many instances (see Chapter 10, particularly Table 10-1 and Fig. 10-2), they are likely to find it increasingly important to be explicit about genetic decisions made on behalf of their patients. As suggested in Chapter 2, for example, hospital discharge protocols in the future will probably include two questions to be answered as part of the discharge procedure: 1. Is the disorder genetic? (yes, no, possibly). 2. Was the patient/family informed of the disorder's genetic nature? (yes, no, deferred). Moreover, if genetic advice is given, it is prudent to keep a written record of it. Short of a formal protocol, however, physicians and other clinicians will more often have occasion to make explicit decisions when a genetic or congenital disorder is recognized. This is true because genetic and congenital disorders account for an increasing proportion of patient problems and because the patients and families are more frequently asking explicit questions about recurrence. Are the diagnosis and genetics as clear as possible? Have genetic heterogeneity, phenocopy, new mutation, variable expression, and so on been taken into account? Has the communication of the disposition to the patient/family been as clear as possible? The clinician will have to decide whether or not to share responsibility for these matters with a geneticist or genetics team. In either case the clinician, and especially the primary physician, is the critical person in implementing the genetic approach to human disease on a widespread and routine basis.

Nurses

Nurses can play key roles in the genetic counseling process, by virtue of either specialized training (e.g., nurse-geneticists (1), pediatric-nurse-practitioners) or their close involvement with patients in certain settings (public health nursing, newborn nursery, doctors' offices, etc). Our experiences in the Colorado-Wyoming Regional Genetic Counseling Program (CW-RGCP) have borne this out. First, the program's *nurse-geneticist* served as a role-model for the many community nurses we worked with (1,2,3). Her clinical nursing background and specialty training made her uniquely effective in counseling patients and gave her the necessary credentials for providing training and educational experiences to other nurses. Second, public health nurses served as local clinic coordinators, referred many patients (in some areas as much as 40% of the patient load), participated in the genetic evaluation by constructing pedigrees and gathering other data, and provided followup visits to many families in their homes. Their referrals showed that the public health nurse may be the only effective link many rural families have with the health care system. Thus, they are a necessary part of any effort to provide genetic services to a population spread over a large area. Third, *office nurses and nursery nurses* were important initiators of referrals through the attending physician. Fourth, many nurses are responsible for evaluation and disposition in *health department clinics,* including family planning and well-baby clinics. Fifth, nurses are often the first (and sometimes the only) person asked about the possible adverse consequences of environmental agents during pregnancy.

Because of their unique and expanding roles in health care delivery, all nurses, especially those with the professional capacities noted above, need to have more formal education and training in clinical genetics. Public health nurses, pediatric nurse practitioners, and other specialty nurse clinicians should be familiar with general genetic principles (Chapters 1 and 2) and should be able to draw at least preliminary pedigrees, to obtain relevant preliminary history (Chapter 9), to understand the principles governing the adverse effects of environmental agents (Chapter 7), and to appreciate the psychological aspects of counseling (Chapters 10 and 11).

Child Health Associates and Physician Assistants

Increasing numbers of nonphysician clinical specialists are coming onto the health care scene. These include child health associates (CHA), obstetric associates, physician assistants, and similarly qualified specialists in other areas. Because these persons will ordinarily be functioning at the primary-care level or as an associate of a medical or surgical specialist, it is important that they be able to recognize the indications for genetic evaluations and to understand the principles of genetic counseling, management, and followup (Chapter 10). Our positive experiences with CHA personnel through the CW-RGCP has strengthened these convictions.

Social Workers and Others

Social workers of various types may be involved in genetic health care as formal members of a genetic counseling team (Chapter 10), as a source of referral, or as a participant in followup management with particular regard to financial, domestic, and psychological/emotional problems. Social workers as members of programs for the handicapped, adoption agencies, and family-planning services are especially important genetic referral sources. Persons in such positions should be familiar with the principles and techniques recommended above for nurses. Adoption agency workers should also be familiar with a number of specific disorders, including those listed in Tables 3-8, 4-5, and 5-2.

Among other personnel who could apply knowledge of clinical genetics are all those who work with handicapped children, including special-education teachers, speech therapists, physical therapists, and occupational therapists.

Genetic Associates

There are several graduate school programs in the United States that offer masters-degree training in clinical medical genetics. Graduates of these programs are qualified to become members of genetic counseling teams in various capacities (4). Their inclusion in this chapter reflects the fact that the term "geneticist" traditionally has been reserved for specialists with M.D. or Ph.D.

training. Genetic associates are nonetheless specialists in gathering data as part of the evaluation, participating in the preliminary counseling and, with increasing frequency, providing the definitive counseling in certain types of cases (1,4,5). There is no question as to their effectiveness in enhancing the flexibility, utility, availability, and impact of genetic counseling teams. They are referred to by various names including "genetic counselors," "genetic assistants," and "genetic associates." The first term is undesirable because they do more than counsel; the second is undesirable because they are more than assistants.

I am an advocate for expanding the role of genetic associates in the delivery of genetic health care. Their participation has already allowed for a very important increase in genetic counseling resources. In the future I envision physicians in group practices hiring their own genetic associate to assist them in directly providing genetic services on a local basis. This would be a logical extension of the genetic associate's role in the CW-RGCP, where the associate accompanied the counselor to the local community for the genetic evaluation and counseling. Thus, genetic associates could work directly with local physicians in a given area to initiate genetic evaluations, provide preliminary counseling, and in some cases engage in definitive counseling. Periodic consultation with the medical geneticist in the regional program would ensure necessary backup. Innovative efforts like these will probably be necessary if genetic health care is to be made uniformly available over wide areas, which is in turn necessary if applied medical genetics is to have a full-scale preventive impact.

Cited References

1. Matthews AL, Blu J, Riccardi VM: Expanded roles of the nurse geneticist and genetic associate. *Am J Hum Genet* 27:64a, 1975.

2. Riccardi VM: Regional genetic counseling programs. in Milunsky A (ed): *The Prevention of Genetic Disease and Mental Retardation*. Philadelphia, WB Saunders, 1975, pp 410–421.

3. Riccardi VM, Robinson A: Preventive medicine through genetic counseling: A regional approach. *Prev Med* 4:126, 1975.

4. Fraser FC: Genetic counseling. *Am J Hum Genet* 26:636, 1974 (especially see page 656).

5. Marks JH, Richter ML: The genetic associate: A new health professional. *Am J Publ Hlth* 66:388, 1976.

CHAPTER THIRTEEN

THERAPY OF GENETIC DISORDERS

Introduction

Therapy of genetic disorders is an important topic in this book for the following reasons:

1. The levels of therapy (diagrammed in Fig. 13-1) demonstrate how inborn errors of metabolism can lead to clinical problems.
2. The range of disorders that can be treated indicates on one hand the limited availability of therapy and on the other its importance in genetic counseling when it is available.
3. In addition to the consequences of therapy for the patient, it may influence other aspects of the genetic counseling process.
4. As shown in Fig. 13-1, direct gene manipulation ("genetic engineering") is not part of the current therapy of genetic disease.
5. Contrary to popular opinion, therapy of genetic disease has not altered the genetic makeup of the human race. This partly reflects its limited availability in terms of time and place.

At present, surgery as tertiary prevention is the most common type of therapy for genetic disorders, especially those with congenital onset. The secondary and tertiary prevention consequences of applied pharmacogenetics are also important, and will be dealt with in Chapter 14.

Availability and Applicability of Treatment Modalities for Genetic Disorders

Because so many genetic disorders involve infants and children, it makes sense to emphasize the preventive potential of each therapeutic effort (Table 11-1, p. 218). Since the affected individual is already born, however, we cannot speak of primary prevention. Secondary prevention is aimed at keeping the gene disorder from expressing itself as a significant clinical problem (e.g., effective drug therapy for epilepsy). Tertiary prevention is aimed at minimizing the problem caused by the disorder's expression (e.g., surgical correction of congenital heart disease). Often, when the therapy of genetic disorders is discussed, dietary and biochemical manipulations are considered the most important methods. Although these approaches have a unique importance in clinical genetics, surgery is the most important type of therapy at present. Table 13-1 draws data from Table 2-6 (p. 18) for a list of the diagnostic catagories in which all or some cases have a genetic etiology and secondary or tertiary prevention efforts may be useful (as opposed to mere palliation in Werdnig-Hoffmann disease or Huntington chorea). Thus, of the first 804 referral diagnoses in the CW-RGCP, fewer than 6% were amenable to secondary (nonsurgical) preventive treatment, fewer than 6% were amenable to nonsurgical tertiary preventive therapy, while more than 30% were amenable to surgical therapy (tertiary prevention).

Table 13-1 does not include surgery for secondary prevention, but this approach is possible in some instances, such as surgery for metabolic ends. This may involve organ transplants to supply deficient enzymes [e.g., Fabry disease (1)] or shunts to divert metabolites from the liver, such as portacaval shunts in glycogen storage disease I or homozygous familial hypercholesterolemia (2).

Table 13-2 lists the three basic categories of genetic disorders and the relative importance of surgical and nonsurgical therapy for each type of subgrouping. Surgical therapy is the most important modality for chromosome disorders, single-gene disorders of congenital onset with morphologic aberrations, and polygenic

TABLE 13-1
These Are Disorders from Table 2-6 That Can Be Treated. They Are Categorized According to Type of Therapy and Level of Intervention. The Percentages Represent Their Proportion of the First 804 Referral Diagnoses.

Category	%
I. Secondary prevention (Nonsurgical)	
Seizures	1.7
Diabetes mellitus	1.3
Hyperlipidemia	1.0
Amino acidopathies	0.6
Thyroid dysfunction	0.5
	5.1%
II. Tertiary prevention	
A. Surgical	
Down syndrome	5.9
Cleft lip and/or palate	5.5
Neural tube defect	3.4
Congenital heart disease	3.3
Neurofibromatosis	2.0
Chromosome rearrangement	1.5
Limb malformations (single-gene)	1.4
Brachy-, syn-, polydactyly	1.1
Urinary tract abnormalities	1.1
Hydrocephalus	1.0
Polycystic/dysplastic kidney	1.0
Charcot-Marie-Tooth disease	0.8
Strabismus	0.5
Marfan syndrome	0.4
Nonspecific joint laxity syndrome	0.4
Warkany syndrome	0.3
	30.9%
B. Nonsurgical	
Sex chromosome aneuploidy	1.7
Cystic fibrosis	1.9
Sickle-cell anemia	1.1
Hemophilia	0.8
	5.0%

TABLE 13-2

Relative Importance of Surgical and Nonsurgical Therapy Modalities for the Three Types of Genetic Disorders

Type of Genetic Disorder	*Example*	*Surgical Therapy*	*Nonsurgical Therapy*
1. Chromosomal	Down syndrome	Major	Minor
2. Single-gene			
Congenital			
morphologic	Holt-Oram syndrome	Major	Minor
nonmorphologic	cystic fibrosis	Minor	Major
Noncongenital			
additional insult required			
dietary	phenylketonuria	Minor	Major
drugs, etc.	G-6-PD deficiency	Minor	Major
no additional insult			
early onset	Werdnig-Hoffmann disease, hemophilia, neurofibromatosis	Variable	Variable
late onset	Huntington chorea, adult Gaucher disease	Variable	Variable
3. Polygenic			
Morphologic	pyloric stenosis	Major	Minor
Nonmorphologic	diabetes mellitus	Minor	Major

Current therapy for genetic disorders mainly involves surgery, avoidance of additional environmental insults, or replacement of a metabolic substance (e.g., insulin, coagulation Factor VIII, etc.). *Major* = The major therapeutic consideration; the modality most likely to be useful. *Minor* = The minor therapeutic consideration; the modality less likely to be useful. *Variable* = No generalization possible; see Text.

morphologic anomalies. Surgery has a variable role in single-gene disorders that are noncongenital, depending on whether they produce changes in tissue amenable to surgery, such as organomegaly (Gaucher disease, glycogen storage disease I, etc.) and neoplasia (neurofibromatosis, intestinal polyposis, etc.). Nonsurgical (medical) measures, of course, may also be part of the therapy for all these disorders. Medical therapy is the most important modality for single-gene disorders with nonmorphologic components and noncongenital onset, especially if adverse environmental factors can be specified, such as dietary phenylalanine in phenylketonuria and sulfa drugs in glucose-6-phosphate dehydrogenase (G-6-PD) deficiency. When no environmental insult can be identified, medical therapy may have very limited application. In turn, surgery is sometimes indicated when patients are being treated medically (e.g., hemophilia, hyperlipidemias).

Therapy as a Factor in Genetic Counseling

For many genetic disorders, medical or surgical therapy is only partially successful and may be associated with its own morbidity, including adverse drug responses and surgical complications. Moreover, when treatment methods are available their benefits should be judged in terms of the patient's whole life and the family's involvement. When the possibility of helping other patients with the disorder enters the picture, the investigational elements of the therapy must be made clear.

Unusual methods of treatment or treatment of rare disorders may be important not only for the patient in question, but also for other, more common disorders. For example, treatment of the relatively rare patient with homozygous familial hypercholesterolemia may teach us how to treat the common hyperlipidemias that contribute to atherosclerotic heart disease.

For the family, the availability of treatment will influence their interaction with the genetic counselor and determine the burden of the disorder for them, as well as for the patient. If the survival of an affected child depends on multiple major surgical procedures and/or complicated dietary or drug regimens, plus frequent hospitalization, the parents might wish to avoid having any other children, affected or normal. Whenever therapy is available, most families have a strong impulse to use it, and this enters into their interaction with the geneticist. Again, one can see that there is more to genetic counseling than the assessment and communication of recurrence risks.

Therapy as an Indicator of Biochemical Pathogenesis

The treatment of inborn metabolic disorders is characterized by extreme specificity: precisely how the biochemical defect leads to the clinical problems determines the therapy used. Understanding the role of therapy reflects an understanding of pathogenesis. Figure 13-1 shows the levels at which therapeutic intervention is possible (see also Fig. 6-1). Any given disorder can usually be treated at only one or two of the seven levels.

At *level 1,* therapy would be aimed at altering the structure or availability of the gene in question. This could be done by causing the mutant gene to undergo a "reverse mutation" which would reestablish the normal, or wild-type status. Or a second normal gene could be introduced, as for example one carried into the cell nucleus by a virus (3). There is ample precedence for introducing genes into bacterial cells by viruses. The main limitations here are of two types: 1. How can the change be effected in all cells where it would be needed? and 2. Is it possible that other changes or deleterious genes might be introduced? The approach is still at a very preliminary, purely investigative stage. Because the implications of altering a person's genetic material are so far-reaching,

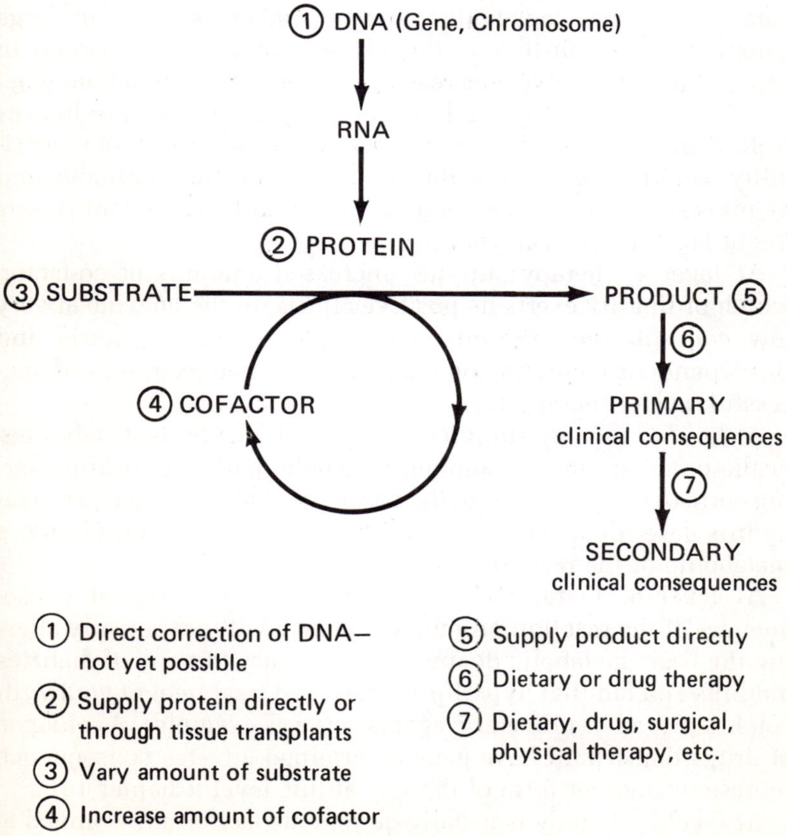

Figure 13–1 Levels of specific therapeutic intervention in heritable biochemical disorders.

research on this "genetic engineering" is being closely scrutinized and monitored.

At *level 2,* therapy would lead to increased production of the protein, decreased breakdown of it, improved function of the protein that is present though inactive, or supplying the protein from a different source. Considerable efforts have been made with respect to the latter—exogenous sources—for a number of diseases. Three well-known examples of replacement therapy are giving insulin to diabetic patients, Factor VIII to hemophilia-A patients, and blood transfusions to sickle-cell anemia patients. Attempts to replace deficient enzymes have been largely confined to lipid storage diseases (4,5), using such materials as serum, white blood cells, and transplanted organs (including kidneys) to supply the enzyme. So far the benefits have been equivocal or too short-lived to be practicable, with the possible exception of renal transplants for patients with Fabry disease (1). The major limitations are the transient effect of the supplied protein and the development of antibodies against it.

At *level 3,* therapy depends on whether the substrate participates in another metabolic pathway and/or is toxic in large amounts. If the substrate or one of its metabolites is toxic (as in PKU), therapy involves decreasing the amount of substrate (e.g., by dietary restrictions) and perhaps supplying the product as well. If the substrate and its metabolites are not toxic, one possibility would be to increase the amount of substrate available and by mass action overcome the block. Vitamin-D-dependent rickets might fall into this category (6).

At *level 4,* therapy supplies increased amounts of co-factor, which ordinarily exerts its positive effects on the enzyme at very low concentration. Pyridoxine-dependent homocystinuria and B_{12}-dependent methylmalonic aciduria are two examples of successful level-4 therapy (7).

At *level 5,* therapy supplies the product, one of its metabolites, or the precursor to it in another metabolic pathway. Administering cortisol to a patient with the adrenogenital syndrome (21-beta-hydroxylase deficiency) exemplifies therapy by supplying a metabolite of the reaction product.

At level 6, therapy minimizes the primary clinical consequences of the reaction product deficiency without actually altering the basic metabolic derangement. Dietary therapy of diabetes mellitus and familial hyperlipidemias, and treatment of gout with colchicine and/or uricosuric agents serve as examples. Avoidance of drugs that would have gene-determined adverse consequence represents another form of therapy at this level (Chapter 14).

At *level 7,* therapy is usually quite nonspecific and is aimed at minimizing the secondary clinical consequences if therapy at the previous levels is ineffective or nonexistent. This includes such procedures as surgery to remove renal and ureteral cystine stones in a patient with cystinuria, aortic surgery in a patient with the Marfan syndrome, and physical therapy for a patient with contractures associated with Charcot-Marie-Tooth disease (peroneal muscular atrophy).

Genetic Engineering and Treatment of Genetic Disorders

The subject of genetic engineering is brought up here only to put it into perspective. As we have seen, it has no place in applied clinical genetics at present. Nor do I believe it will in the near future. This forecast is based on both the social implications and technological limitations of genetic engineering. In addition, the underutilization of currently available genetic counseling resources and the real prospects for primary prevention indicate where priorities should be placed.

Therapy and the Prevalence of Mutant Genes

An argument against vigorous treatment of patients with heritable disorders is the concern that the human gene pool will somehow be weakened by an increase in mutant genes. This argument is untenable: It is incorrect from a genetic standpoint, it disregards the rights of individual persons, and it is disrespectful of the common sense that most people have.

In the absence of specific selection pressures, it takes many generations to alter the proportion of rare mutant genes in the population (8,9). Given the relatively recent advent of genetic therapy, its still nonspecific nature, and the limited populations to which it is available, at this point we need not be concerned about altering the gene pool. Even if therapy were a significant positive pressure, it would involve a very small portion of the world population carrying autosomal recessive mutations. The persons requiring therapy are the homozygotes and they are much less common than heterozygotes. For example, the Caucasian incidence of homozygotes with cystic fibrosis is 1/1600, while the incidence of heterozygotes is 1/20. For X-linked recessive mutations the frequencies of unaffected heterozygotes and affected hemizygotes may be more nearly comparable, depending on whether the disorder is a genetic lethal. With dominant genes, where the heterozygotes are affected, the common sense of affected persons receiving genetic counseling may be as much a factor as therapy in altering the gene frequencies in the population. Patients affected with a high-burden dominant disorder are likely to avoid passing it on if they know it is transmissible and that there may be acceptable alternatives.

For common disorders with important heritable components, such as diabetes mellitus, hypertension, hyperlipidemia, atherosclerotic heart disease, and polygenic malformations, environmental factors play a very large role. As they are identified and avoided, the genetic factors become less important as contributors to their respective disorders.

A further consideration is the contribution of mutant genes to the population, not by individuals affected with heritable disorders, but by persons who have been exposed to mutagenic agents. In this group special attention might be paid, for example, to individuals capable of procreating who have been treated for cancer, since many anticancer agents are known to be mutagens and/or oncogens (10–15).

Conclusion

Therapy for genetic disorders is usually nonspecific; it is limited in scope; and most often it involves tertiary prevention. When available, it is an important aspect of genetic counseling on many levels. Researchers are trying to bring therapy closer to the mutant gene, but much work remains to be done. Genetic engineering is not now a treatment method and concerns for adverse alterations of the human gene pool are untenable. As more and more treatable genetic disorders are being recognized, genetic counseling and therapy resources must be made available to affected patients and their families.

Cited References

1. Clarke JTR, Guttmann RD, Wolfe LS, et al: Enzyme replacement therapy by renal allotransplantation in Fabry's disease. *New Engl J Med* 287:1215, 1972.

2. Stein EA, Pettifor J, Mieny C, et al: Portocaval shunt in four patients with homozygous hypercholesterolemia. *Lancet* 1:832, 1975.

3. Cohen SN: Gene manipulation. *New Engl J Med* 294:883, 1976.

4. Brady RO: The lipid storage diseases: New concepts and control. *Ann Intern Med* 82:257, 1975.

5. Brady RO, Pentcheo PG, Gal AE: Investigations in enzyme replacement therapy in lipid storage diseases. *Federation Proc.* 34:1310, 1975.

6. Fraser D, Kooh SW, Kind HP, et al: Pathogenesis of hereditary vitamin-D-dependent rickets. An inborn error of vitamin D metabolism involving defective conversion of 25-hydroxyvitamin D to 1-alpha, 25-dihydroxy-vitamin D. *New Engl J Med* 289:817, 1973.

7. Ampola MG, Mahoney MJ, Nakamura E, et al: Prenatal therapy of a patient with vitamin B_{12}-responsive methylmalonic acidemia. *New Engl J Med* 2931:313, 1975.

8. Stern C: Selection in civilization. in *Principles of Human Genetics,* (3rd ed.). WH Freeman, San Francisco, 1973, pp 754–798.

9. Fraser GR: The implications of prevention and treatment of inherited disease for the genetic future of mankind. *J Genet Hum* 20:185, 1972.

10. Lewis EB: Possible genetic consequences of irradiation of tumors in childhood. *Radiol* 114:147, 1975.

11. Li FP, Stone R: Survivors of cancer in childhood. *Ann Intern Med* 84:551, 1976.

12. Harris CC: The carcinogenicity of anticancer drugs: A hazard in man. *Cancer* 37:1014, 1976.

13. Favus MJ, Schneider AB, Stachura ME, et al: Thyroid cancer occurring as a late consequence of head and neck irradiation. *New Engl J Med* 294:1019, 1976.

14. Trosko JE: The ethics of "conquer cancer now." Presented at Am Inst Biol Sciences, New Orleans, LA, 4 June 1976.

15. Huberman E: Correlation between mutagenicity and carcinogenicity. *Mutat Res* 29:285, 1975.

General References

A. Buckley RH, Whisnant JK, Schiff RI, et al: Correction of severe combined immunodeficiency by fetal liver cells. *New Engl J Med* 294:1076, 1976.

B. Desnick RJ, Krivit W, Fiddler MR: Enzyme therapy in genetic diseases: Progress, principles and prospects. in Milunsky A (ed) *The Prevention of Genetic Disease and Mental Retardation*. Philadelphia, WB Saunders, 1975, pp 317–342.

C. Desnick RJ, Thorpe SR, Fiddler MR: Toward enzyme therapy for lyposomal storage disease. *Physiol Reviews*, 56:57, 1976.

D. Hsia YE: Treatment in genetic diseases. in Milunsky A (ed). *The Prevention of Genetic Diseases and Mental Retardation*, Philadelphia, WB Saunders, 1975, pp 277–305.

E. Lorber J: Selective treatment of myelomeningocele: To treat or not to treat. *Pediatrics* 53:307, 1974.

F. Milunsky A: *The Prevention of Genetic Disease and Mental Retardation*. Philadelphia, WB Saunders, 1975.

G. Shenasky JH II, Whelchel JD: Renal transplantation in Prune Belly syndrome. *J Urol* 115:112, 1976.

CHAPTER FOURTEEN

PHARMACOGENETICS

Introduction

Drugs are capable of doing things *to* patients, not only *for* them. Often the explanation for patient–drug interactions has a genetic basis. Pharmacogenetics deals with the effects of drugs that are explicable, and therefore predictable, on the basis of the patient's genetic makeup. When we consider the large number of patients using drugs, and the relatively large number of drugs used per patient, pharmacogenetics takes on significance for health care workers, especially those who prescribe and dispense drugs. An appreciation of the principles of pharmacogenetics can help prevent many adverse drug responses and, conversely, it can directly enhance the therapeutic effort (i.e., the use of allopurinol for certain forms of gout). The application of those principles is an important way of implementing noncounseling genetic care.

Drug responses can be explained on a genetic basis in terms of *mechanisms* (i.e., site and type of patient–drug interaction) or in terms of the *clinical context* in which they occur. To demonstrate these concepts this chapter will review a number of well known genetic bases for distinctive, usually adverse drug reactions. With regard to mechanisms, there are four general areas of concern:

1. *Drug metabolism defect:* Normal doses lead to abnormally high or low blood levels, which cause excessive or diminished responses, respectively.
2. *Tissue metabolism defect:* Normal doses lead to normal blood levels, but the drug is processed abnormally at the target tissue.
3. *Receptor-site aberrations:* Normal blood levels of the drug, but the drug's interaction with its receptor is impaired.
4. *Abnormal anatomy:* Distortions of tissue or organ structure.

An awareness of the clinical context in which adverse drug responses occur serves as a starting place for minimizing many of these responses. The point of distinguishing them is to emphasize how they may be recognized in specific clinical settings. The first two types of traits noted below (1a, 1b) are essentially defined in terms of the adverse drug response. Thus, inquiries about them must specifically ask about the drugs involved and responses to taking them. The second two types of traits (2a, 2b) are defined by multiple features, one of which includes a distinctive drug response, either deleterious (2a) or salutory (2b). Inquiries about these disorders thus involve questions about certain drugs and responses to them, as well as questions about features of the disorder unrelated to drug ingestion. The four types of clinical settings may be classified as follows:

1. Mendelian traits characterized by a variant pattern of response to drugs:
   (a) The adverse drug response is the only clinical manifestation of the trait (e.g., succinyl choline sensitivity).
   (b) The adverse response is the usual, though not the only, clinical manifestation of the trait (e.g., hemolysis in G6PD deficiency).
2. Genetic disorders with altered drug sensitivity:
   (a) Though clinically important, the adverse response is only *one of several deleterious manifestations* of the trait (i.e., exacerbation of acute intermittent porphyria by barbituates).
   (b) The genetic disorder itself is the *reason for administering* specific drugs (e.g., cholestyramine for hypercholesterolemia, allopurinol and/or other drugs for gout).

ABNORMAL RESPONSES—CLINICAL CHARACTERIZATION

Variant Drug Response as the Trait's Only Feature

In this group of disorders there are no associated clinical features except perhaps a positive family history. Three important examples will be discussed here.

Succinyl choline sensitivity produces an exaggerated and prolonged muscle paralysis after administration of a standard dose of succinyl choline (SC), which is given for muscle relaxation during surgery (1). The primary abnormality is in pseudocholinesterase, an enzyme that acts on SC and similar compounds, converting them to inactive compounds. The defective enzyme has a decreased ability to interact with SC, so that a given dose takes longer to inactivate and thereby has an exaggerated effect. This autosomal recessive disorder can be tested for by gauging the inhibitory effect of dibucaine on the enzyme reaction. The derived "dibucaine number" represents the percentage inhibition of the enzyme and is clinically useful for assessing at-risk persons. At or above 73% inhibition indicates the normal homozygous state, 45–70% the heterozygous state, and less than 20% the affected homozygote. About 1 in 3000 of the general population is an affected homozygote.

The *N-acetyltransferase variant* is important because the enzyme N-acetyltransferase is involved in detoxifying a number of commonly used drugs, including isoniazid, phenelzine, hydralazine, dapsone, and sulfadimidine (1,2). The decreased rate of detoxification can increase the likelihood of drug side effects and direct toxicity—e.g., neuropathy from isoniazid. This autosomal recessive trait is unusual in that what we call the mutant phenotype and genotype are more common than the "normal" or wild-type among whites and blacks (60%), although among Orientals its frequency varies from 5–20%. The slow acetylators can be detected by a clearance test using sulfamethazine, but because the test is not widely available, reliance is placed on careful clinical assessment and monitoring of blood levels.

Warfarin resistance involves a defect at the intracellular binding site for warfarin anticoagulants (3). This autosomal dominant trait is rare, but warfarin use is so common that the variant should be known to all who prescribe and dispense it. Blood and tissue levels are proportionate to dosage, but with the trait a therapeutic effect requires about 145 mg per day as opposed to the normal dose of 5–10 mg per day. There is an associated 20-fold increase in sensitivity to the effects of vitamin K in promoting coagulation.

Variant Drug Response as the Usual Feature, Though Not the Sole One

There may be nonspecific associated clinical features to forecast the adverse response before drug administration.

Malignant hyperthermia is an autosomal dominant trait that has a frequency of about 1/10,000 (4). The adverse response involves the development of extreme fevers, as high as 41–42°C, associated with muscle rigidity, and cardiovascular and neurologic complications. The offending agents have not been identified specifically but they are part of the general anesthesia regimen. There may be evidence of muscular weakness or other muscle problems in the patient's history and perhaps elevations of serum creatine phosphokinase levels in affected patients (5). At present, however, the family history is the critical clinical feature: inquiries about adverse reactions to anesthesia (or surgery) among family members must be part of routine preanesthetic evaluations.

Glucose-6-phosphate dehydrogenase (G6PD) deficiency is an X-linked recessive trait that is very common among blacks, as well as whites of Mediterranean origin, especially Sardinians, Sicilians, and Greeks (6). The enzyme is critical to energy metabolism in red blood cells, especially in supplying the co-factor nicotinamide adenine dinucleotide phosphate in reduced form (NADPH). When NADPH is diminished, red blood cells become susceptible to destruction (hemolysis) by a wide variety of drugs, including chloramphenicol, sulfonamides, dapsone, primaquine, and acetanilide (7). The hemolysis may result not only in anemia, but perhaps also secondary renal damage, which may cause death. The at-risk persons are hemizygous males, and females who are either homozygous, reflecting the mutant gene's high frequency, or functionally hemizygous by virtue of the Lyon principle (Chapter 4). These persons may show chronic low-grade anemia and can be detected with a G6PD screening test or enzyme electrophoresis. The drug-caused problems secondary to G6PD deficiency are neither mild nor rare, and they can be prevented by good personal medical histories, family histories, and screening tests in at-risk populations.

Penicillin allergy. This abnormal response is immunologically mediated and consequences may range from a nuisance rash and itching to death in anaphylactic shock. Although clearly a familial trait, its genetic nature has proved elusive. The two most likely possibilities are polygenic or autosomal dominant with variable expression. In any case, first- and second-degree relatives of an affected person are at increased risk compared to the general population. Before parenteral administration of penicillin a family history is mandatory.

Genetic Disorders in Which an Adverse Drug Response is Only One of Several Deleterious Manifestations

There are many disorders in this category, and we will consider three examples. Ordinarily, but not always, the trait is otherwise apparent and the drug response aggravates the clinical course.

Acute intermittent porphyria (AIP), an autosomal dominant trait, is the prototype of this group. The disorder can be precipitated or exacerbated by small doses of many drugs (8). The barbituates are especially important because they are often given for symptoms of AIP. The mechanisms by which a drug worsens the disease is not known, but death or severe disability may ensue. Personal health history, family history, and screening tests (e.g., Watson-Schwartz test) can minimize the drug-induced problems. Other types of porphyria (e.g., hereditary coproporphyria) may also be exacerbated by drugs.

Hypertrophic subaortic stenosis (HSS) is an autosomal dominant trait that occurs alone (9) or with other disorders, such as the Noonan syndrome. Cardiovascular disturbance may result, including congestive heart failure for which the usual treatment with digitalis preparations is absolutely contraindicated. These preparations aggravate the out-flow obstruction with severe consequences, including death. A positive family history for either HSS or adverse response to digitalis preparations indicates at-risk patients.

Glucocorticoid-induced increase in intra-ocular pressure in glaucoma due to trabecular defects is well known (10). This adverse response is clinically an autosomal dominant trait: the adverse effects of topical steroids are seen in the heterozygote. However, the effect is even more pronounced in patients homozygous for the mutant gene. The anatomic basis for glaucoma and the family history should be known before topical glucocorticoids are applied to the glaucomatous eye.

Genetic Disorders as Indications for Specific Drug Therapies

In some heritable disorders, the trait itself provides a rationale for specific therapy (Chapter 13). These include corticosteroids for the adrenogenital syndrome, penicillamine for Wilson disease, cholestyramine for familial hypercholesterolemia, allopurinol for gout, diiodohydroxyquin for acrodermatitis enteropathica.

MECHANISMS OF ABNORMAL RESPONSES

Because the clinical context of drug ingestion may not readily forecast adverse responses (2a, 2b above), it is important to emphasize mechanisms of adverse drug responses. By appreciating them it may be easier to probe patient histories for indicators of genetically based drug catastrophies.

With *drug metabolism defects* usual doses lead to abnormally high or low blood or tissue levels, as in succinyl choline sensitivity and the N-acetyltransferase variant.

With *tissue metabolism defects* normal doses lead to normal blood levels, but aberrant tissue responses. This occurs in malignant hyperthermia, G6PD deficiency, acute intermittent prophyria, and methemoglobin reductase deficiency. The last of these is present in the heterozygous form in about 1% of the general population (11). In the homozygous state it accounts for a rare type of autosomal recessive methemoglobinemia. Cyanosis of some severity may develop in heterozygotes in response to certain drugs, including dapsone, chloroquine, and primaquine.

Receptor-site aberration has been specified for only one disorder—warfarin resistance. Nonetheless, this will probably turn out to be a relatively common mechanism of adverse drug responses.

Anatomic defects may lead to adverse responses without involving abnormal metabolism in any sense: the tissue or organ responds abnormally as a function of its deformed structure. Two examples given above are the HSS response to digitalis preparations and increased intra-ocular pressure due to topical steroid treatment of glaucoma due to trabecular defects.

Adverse drug responses involving mutagenesis, oncogenesis, chromosome damage, teratogenesis, and very early abortion were discussed in Chapter 7.

Conclusion

Two general points derive from the foregoing: 1. There is indeed a genetic component to many drug responses; 2. A drug history to minimize adverse reactions must therefore include inquiries about other family members. The limited number of examples noted here is intended to facilitate implementation of these generalizations into routine clinical situations. For a more complete itemization and depiction of such disorders the reader is referred to the "General References" noted below.

Cited References

1. LaDu BN: The isoniazid and pseudo-cholinesterase polymorphisms. *Fed Proc* 31:1276, 1972.

2. Evans DAP, Manley KA, McKusick VA: Genetic control of isoniazid metabolism in man. *Br Med J* 2:485, 1960.

3. O'Reilly RA: The second reported kindred with hereditary resistance to oral anti-coagulant drugs. *New Engl J Med* 282:1448, 1970.

4. King JO, Denborough MA, Zapf PW: Inheritance of malignant hyperpyrexia. *Lancet* 1:365, 1972.

5. Isaacs H, Heffron JJA, Badenhorst M: Predictive tests for malignant hyperpyrexia. *Br J Anaesth* 47:1075, 1975.

6. Motulsky AG, Yoshida A, Stamatoyannopoulous G: Variants of glucose-6-phosphate dehydrogenase. *Ann NY Acad Sci* 179:636, 1971.

7. Fraser IM, Tilton BE, Vessell ES: Effects of some metabolites of hemolytic drugs on young and old, normal and G6PD-deficient human erythrocytes. *Ann NY Acad Sci* 179:644, 1971.

8. Tschudy DP, Valsamis M, Magnussen CR: Acute intermittent porphyria: Clinical and selected research aspects. *Ann Intern Med* 83:851, 1975.

9. Clark CE, Henry WL, Epstein SE: Familial prevalence and genetic transmission of idiopathic hypertrophic subaortic stenosis. *New Engl J Med* 289:709, 1973.

10. Armaly MF: Genetic factors related to glaucoma. *Ann NY Acad Sci* 151:861, 1968.

11. Bloom GE, Zarkowsky HS: Heterogeneity of the enzyme defect in congenital methemoglobenemia. *New Engl J Med* 281:919, 1970.

General References

A. Vessell ES: Advances in pharmacogenetics. *Progr Med Genet* 9:291, 1973.

B. World Health Organ: Pharmocogenetics. *World Health Organization Technical Reports Series No. 524*, World Health Organization, Geneva, 1973.

C. Yeshurun D, Gotto AM Jr: Drug treatment of hyperlipidemia. *Am J Med* 60:379, 1976.

CHAPTER FIFTEEN

ETHICAL, MORAL, AND LEGAL ASPECTS OF CLINICAL GENETICS

Introduction

The many ethical, moral, and legal controversies in clinical genetics have actually made genetics more a part of health care in general. Although genetic disorders and techniques like amniocentesis and screening have focused attention on certain problems, the heightened awareness has led to a greater demand for genetic services. In this chapter we will examine some of the concerns over the potential and actual impact of applied medical genetics and some of the societal implications of genetic counseling and its techniques.

"Quality of Life"

The phrase "quality of life" refers to the fact that there is more to human existence than merely being alive, that there is a level of capability that adds a special dimension, or quality, to the individual's life. If that capability is not attainable the individual's "quality of life" is thereby impoverished. Although there are varying degrees of impoverishment, at times it is actually or potentially so great that decisions made on behalf of the affected individual may be predicated on the impoverished quality of life. In the process of genetic counseling, implicit assumptions about quality of life become translated into explicit decisions on behalf of affected or at-risk individuals. These decisions, made many times each day, involve such issues as artificial insemination by donor (AID), prenatal diagnosis, selective abortion, and management of malformed and retarded children. Thus, prevailing attitudes about quality of life have an important influence on decisions made in genetic counseling. In addition, several recent developments in biology, medicine, and law have dramatically increased the number of occasions when decisions based on quality of life need to be made. Some of the recent developments which have made such decisions part of our routine daily lives include the following:

1. There are experimental systems which treat the biological results of *in vitro* fertilization differently from the results of natural fertilization, and which treat a fetus about to be aborted differently from one in a pregnancy that will continue. A conceptus resulting from *in vitro* fertilization and a fetus about to be aborted are given a unique significance: Each is part of an endeavor that is contrary to its own well-being. Through such experimental approaches we indicate that life has different levels of meaning, and implicit to this may be notions about quality of life. For example, some may say that the fetus about to be aborted can never attain the same quality of life potentially available to the fetus to be delivered and it can thus be treated without concern for impairing its ultimate quality of life. Our attitudes toward each of the fetuses is determined, at least in part, by presumptions about quality of life.
2. Prenatal diagnosis enables geneticists to forecast quality of life from the results of a laboratory test, and this forecast may then become the basis for terminating the pregnancy. Knowing that a fetus has three number 18 chromosomes comes to nought unless it indicates something about the bearer of those three chromosomes—unless it portends something about the expected quality of life.
3. Legal contests have arisen over whether a given quality of life is grounds for cessation of extraordinary life-support and whether a malformed child is entitled to compensation for an impoverished quality of life (1–4). In the latter case, it is a matter of deciding whether a given quality of life is to be considered so undesirable that terminating the pregnancy, were the choice available, would even be reasonable from the affected child's point of view.

Decisions resulting from genetic counseling (e.g., AID, prenatal diagnosis, selective abortion) are moving medicine beyond its traditional boundaries and are themselves subject to many nonmedical influences. Because clinical genetics provides this interface between medical and social issues, and because decisions relevant to these issues are made daily, with immediate application by counseled families, it is very important that health care workers involved in genetic counseling be aware of their own personal attitudes toward quality of life. For example, does the person assume that an 18-week pregnancy with a demonstrated trisomy-18 fetus would be terminated? Or, does the counselor feel that it is somehow wrong for a couple to take the 1 in 4 risk for having another child with cystic fibrosis? Genetic health care providers, including primary care practitioners, must recognize their thoughts and feelings about these issues so that they don't interfere with objective counseling. It is especially important that one person's attitudes do not deliberately or inadvertently preclude another person from the possibility of utilizing available alternatives.

Informed Consent and the Rights of Patients

The issues of informed consent and patient rights are relevant at all levels of medical genetics, from investigation to clinical practice. In several areas of applied genetics they are particularly important.

1. The right of the patient and/or family to a genetic referral and adequate risk apprisal regardless of the biases of their primary physician.
2. The right to privacy in the context of the genetic counselor's obligation to inform all at-risk family members, perhaps contrary to the consultand's wishes.
3. The efficacy and consequences of genetic screening programs.
4. The use and limitations of prenatal diagnosis, especially through amniocentesis.
5. Sterilization of the severely handicapped.
6. Limiting management alternatives as a consequence of making a specific genetic diagnosis.
7. Diagnostic and therapeutic measures of an investigative nature that require the patient's informed consent.

The first point, the right to referral and apprisal, has been discussed in depth already (Chapter 1, 2, and 9). It is a right that has a legal as well as an ethical basis (5).

The second point, right to privacy, is full of complexities, some unique to clinical genetics. Basically we are concerned with the implicit or explicit wishes of the patient or family that medical–genetic data not be divulged except in accord with traditional medical–legal guidelines. First, there is the chance of violation of privacy if the family is included in a large genetics registry. Among the purpose of such registries are the maximal utilization of available data on rare disorders for the care of other families, and the ability to resume contact with families in the event of improvements in therapy, carrier detection, diagnosis in general, and especially prenatal diagnosis. With strict monitoring and built-in safeguards the risks of a trespass of privacy are far outweighed by the potential benefit to affected families. Second, there is a dilemma about data on family members recorded in the evaluation, including the pedigree: if the proband's complete records are released in traditional ways, should certain data from the pedigree be deleted first since, strictly speaking, they are part of someone else's medical history? I think not, unless the patient or a relative requested the deletion. However, this implies that there is something about the genetic counseling situation which transcends the usual restrictions: the disorder is not merely an individual but a *familial* problem, and the counselor is necessarily involved with family members. Third, since the geneticist's obligation is not only to the proband, but also to other family members at risk for either developing the disorder or having affected offspring, he must be able to fulfill his obligation to all at-risk family members, even if the proband or parents state their wishes to the contrary. A family member probably does not have a right to deny other at-risk family members knowledge of their risk. There is no legal precedent for this, however.

The third point, regarding genetic screening, has been examined in depth elsewhere (6). It is enough to point out here that, unless the screening program has built-in mechanisms to directly inform the individual or family or their primary physician, the identification of affected persons or carriers may be more harmful than beneficial.

The fourth point, dealing with prenatal diagnosis, has several aspects worth emphasizing in this discussion of ethics. First, the women or couple for whom the procedure is indicated must be fully informed as to its availability, the potential positive and negative consequences, the potential hazards, and the limitations. There are two corollaries: First, it must be available regardless of the biases of the primary physician, and involvement with a specialized prenatal diagnosis genetics program is warranted in all cases. Second, it is not yet clear what the long-term impact of prenatal diagnosis and the prospect of pregnancy termination will be on the relationship between parents and child if the pregnancy is not terminated. Although short-term followup has not revealed adverse consequences (7), we do not know about the long-term effects on psychosocial development. Third, the use of amniocentesis has been restricted to clear-cut medical indications; the technique has not been used for such purposes as sex selection for domestic or social reasons. Up to now, this restriction has largely reflected limited technological resources. When those limitations are overcome, the situation may change. Fourth, in the course of prenatal diagnosis certain decisions are made solely from laboratory data and *the source of the data (the fetus) is not considered a patient*. The pregnant woman is the patient. This is a thorny dilemma, unique in medicine.

The fifth point, sterilization of severely handicapped children frequently comes up during genetic counseling. This is another difficult problem, which can be argued from several viewpoints.

The genetic counseling disposition may add one or two additional considerations, including expected natural sterility of the patient (e.g., a male with the Down syndrome) or the possibility of transmitting the same defect to offspring. However, legal action is necessary to have retarded or otherwise severely handicapped persons sterilized.

The sixth point, limiting management after making a specific genetic diagnosis in a newborn infant, directly involves assumptions about quality of life. The example of a newborn infant suspected of having trisomy 13 (Patau syndrome) can be cited again. Before instituting any extraordinary therapy, it is possible to do a chromosome analysis on a bone marrow aspirate and demonstrate the chromosome abnormality in about 5 hours. If trisomy 13 is documented, the very poor prognosis might lead to a decision to minimize therapeutic efforts, and especially avoid extraordinary procedures. The same tragic situation may arise when children are born with trisomy 18 (Edwards syndrome), triploidy, and the Down syndrome in some cases. The clinical geneticist's role is to demonstrate the chromosome abnormality and clarify its significance. Decisions based on those data are made by the parents with assistance from all professionals with whom they consult.

The seventh point, regarding informed consent, is important in all branches of medicine. Areas of special concern in clinical genetics include prenatal diagnosis, situations in which diagnosis and/or management are entirely or partially investigational, and sterilization of mentally retarded patients. For prenatal diagnosis, informed consent means ensuring that the pregnant woman or couple understands fully the indications, consequences, limitations, and risks of the procedures involved. Since many genetic disorders are rare and their pathogenesis incompletely known, their evaluation and therapy often are not well established and may be investigational. A special effort must be made to clarify to the patient and family what procedures are investigational and what the purpose of the investigation is. Informed consent to sterilization frequently is discussed in genetic counseling sessions, and the main issue is who can give it—the patient, parents, or legal guardian?

Moral, Ethical, and Legal Aspects of Specific Types of Programs

There is a difference between 1. an ethical discussion of genetic counseling, and public education about genetics and 2. a discussion of formal programs designed to implement these approaches on a large scale. With formal programs, especially if they are government sponsored, other elements include such things as compulsory participation and the implication that social institutions have a right to determine or influence genetic makeup.

The various issues revolving around *genetic screening programs* have served to focus public attention on medical genetics (8–12). There seems to be little controversy over situations where the following criteria are met: the at-risk population is defined, the test is simple and inexpensive, and there are positive therapeutic consequences. Screening of the newborn for patients afflicted with phenylketonuria (PKU) or galactosemia falls into this category. Carrier identification for autosomal recessive disorders whose prevention requires prenatal diagnosis, AID, or contraception (e.g., Tay-Sachs disease) is not without controversy. Criticisms have focused on the balance between the cost of the screening and the cost of care for an affected child, possible "excessive anxiety" on the part of identified carriers, and incomplete cooperation on the part of at-risk persons. These concerns in turn indicate an underlying problem: program planners, and often critics as well, sooner or later begin to judge the program in terms of the consumer's compliance regardless of the consumer's own feelings (13). This eventually leads to the question of whether an at-risk couple has a right not to comply since the cost of caring for the affected child they may have will probably be borne by public agencies. The most widely accepted view is that screening should be routinely available, but that use of it and of the results should be completely voluntary. In this regard, and in other ways, Tay-Sachs screening programs are important models for similar programs that, it is hoped, will soon be available for such disorders as cystic fibrosis.

The controversies surrounding screening programs for sickle-cell anemia and sickle trait are many and complicated (14,15). The main issues have been the inability to alter the course of the disease or utilize prenatal diagnosis routinely once affected persons or carriers, respectively, have been identified. A suspicion held by some that racism may be involved has also been a factor. Also, many carriers have been discriminated against socially and in terms of employment and insurance, even though the sickle trait is not a pathologic condition. However, in the midst of all the controversy about sickle-cell screening, the common-sense reasons for a given couple knowing their carrier status is often overlooked. Thus it is important that we distinguish between genetic techniques and programs to implement them.

Sex chromosome screening programs in North America and Western Europe have focused considerable attention on medical, legal, and social aspects of a person's genetic makeup. One point of concern has been the extent to which a person's chromosome constitution determines his behavior and legal liabilities. Another is whether a person's chromosome constitution is better left unknown without a specific and compelling medical reason to the contrary. Apart from two brief comments here, the interested reader is referred to a thorough review of this matter by Hamerton (16). First, considering the relatively common occurrence of sex chromosome aneuploidy, the reasonable suspicion that it may be deleterious (17), and the lack of substantial prospective data, the need for well-run sex chromosome screening programs seems clear. Second, the argument that premorbid awareness of the aneuploidy will itself contribute to morbidity disregards the common sense of most people. It is unreasonable to withhold genetic information or the opportunity to obtain it from the individual in question based solely on presumptions as to how it will be used by him or her.

Programs designed to provide *prenatal diagnosis* raise several questions. Who should fund the programs? Are they equally accessible to all who could or would use them? Is amniocentesis justifiable unless the couple first consents to abortion if findings are positive? Should geneticists alone determine the policies of a prenatal diagnosis program? Can a health care worker withhold referral to a prenatal diagnosis program because of his or her aversion to abortion? These and many other questions are raised by the technologic feasibility of large-scale prenatal diagnosis programs (18,19). Technical considerations have been discussed in Chapter 8 and reviewed elsewhere (20,21). Four points will be made here: 1. Prenatal diagnosis does afford an effective means for minimizing the number of children born with certain common, as well as rare, genetic disorders. 2. To be uniformly available to widespread populations, prenatal diagnosis programs must be coupled with active outreach clinics and education efforts, which are most efficiently funded by states (22,23). 3. Prenatal diagnosis is a means to obtain information, and therapeutic abortion is not a necessary consequence of positive findings. 4. Referral for prenatal diagnosis must be determined by specific indications, not by the attitudes of the referral source.

Legal Aspects of Genetic Counseling and its Techniques

Many professional and lay publications have begun to deal with legal aspects of genetic counseling (24,25). In this section we will consider four issues that have been recurrent themes throughout this book:

1. Does genetic counseling constitute the practice of medicine?
2. Do the obligations of a medical geneticist transcend certain traditional restrictions on physician prerogatives?

3. What constitutes adequate communication of the genetic counseling information to the consultand?
4. Who has responsibility for making genetic services available to all who need them?

It should be clear by now that *genetic counseling is an element of ordinary medical practice,* and this means any licensed physician can provide genetic counseling. But what about nonphysicians providing genetic counseling? There is no question that many nonphysicians can provide excellent counseling, including registered nurses, genetic associates, social workers, and those with Ph.D. training. These multiple levels of genetic health care expertise are important to recognize in planning and legislation for the use of genetic health care resources (26). The main point is to avoid declarations that genetic counseling is, or is not, solely a medical domain. To say that it is may exclude qualified nonphysicians from providing genetic health care. To suggest that it falls outside the bounds of ordinary medicine would so circumscribe genetic counseling as to risk amputating it. At least for the present and near future, we should consider genetic counseling a form of health care that is an extension of medicine governed by the same general guidelines. Responsibility for many components of counseling ultimately fall to a physician but along the way many nonphysicians are critical contributors.

The primary responsibilities of a geneticist are to establish a diagnosis and to communicate its meaning, to effectively provide information to at-risk persons so that they may use it in making rational decisions about childbearing (27). In this activity there are four elements we should touch on here: 1. The importance of consensus diagnosis. 2. The legal consequences of genetic counseling communication. 3. The responsibility for initiating or reestablishing contact with at-risk persons. 4. The geneticist's obligation to inform at-risk persons versus the consultand's claim to privacy.

1. The importance of consensus diagnosis has already been discussed. It is brought up again here to emphasize the consequences of misdiagnosis. If the basis for the information to be communicated (the diagnosis) is incorrect, the information may be erroneous. Not only may decisions about conception and termination of pregnancy be a consequence of the communication, but also retrospective assignment of liability. Increasingly, geneticists are asked to give opinions about possible causation of birth defects in the course of litigation. The need for consensus about an accurate diagnosis or its absence is obviously critical. It is doubtful that any geneticist can accurately diagnose every case referred to him.

2. Even at a minimum, genetic counseling involves a written

account of the evaluation and counseling, for the patient's records as well as the immediate use of the patient or family. The language must be understood by the family and yet be precise enough about technical distinctions so as not to imply associations that are not actually known. For example, an organ or tissue may be smaller than normal because of hypoplasia or atrophy; if the mechanism is not known, only a description of the smallness is warranted.

3. What can a person or family expect with regard to being provided genetic counseling and information about significant new developments? Although it is clear that all at-risk families have a moral and a legal right to genetic counseling it is less clear, especially in a legal sense, whether a counselor is implicitly obligated to initiate contact with an individual or to reestablish contact with a family in light of new developments which may alter the genetic counseling alternatives. That is, if a counselor is aware of new knowledge relevant to a family, and it is likely that no one else could give them the information, is he implicitly obligated to seek out the family and so inform them? The adequate staffing of clinical genetics units and the use of computer registries are specially pertinent to this issue.

4. The concern for contacting other family members while respecting the consultand's and/or proband's privacy is a more complicated problem. I believe that a geneticist has a responsibility to see that at-risk persons recognize they are at-risk. This is one implicit purpose of a family history—to identify at-risk persons for future counseling. Usually there is no obstacle, but occasionally the consultand or proband forbids contacting other family members. The legal aspects of this situation are unclear. The geneticist's tacit and intrinsic obligation to inform other family members of their risks has not been established in the courts.

Adequate communication involves both spoken and written exchanges between the genetic counselor and the proband, parents, or guardians. It must be clear when the exchanges were made and what they consisted of, with special attention to technical concepts, emotional factors, and the fact that decisions may have to be made without the benefit of precedent. Ideally, a copy of the counseling report is sent to the patient or family, but sometimes a summary letter is preferable. In the long run, the content of genetic counseling documents will assume increasing legal significance.

The *availability of specialized genetic services* is relatively limited at present; most resources are located in major medical centers in large cities. Yet efforts to enlarge the areas served, like those described in Chapter 16, require minimal expenditures. These cannot be accounted for by fee-for-service and would best be funded by states (22,23). States unwilling to subsidize adequate outreach programs will be ignoring both their citizens' right to this type of health care and the opportunity to decrease other health care costs (e.g., institutionalization). State-supported programs of this type, however, should be voluntary at all levels and include training and educational components as well.

Cited References

1. Capron AM: Informed decision making in genetic counseling: A dissent to the "wrongful life" debate. *Indiana Law J* 48:581, 1973.

2. Carter G: Legal responses and the right to compensation. *Br Med Bull* 32:89, 1976.

3. Robertson JA, Frost N: Passive euthanasia of defective newborn infants: Legal considerations. *J Pediatr* 88:883, 1976.

4. Waldman AM: Medical ethics and the hopelessly ill child. *J Pediatr* 88:890, 1976.

5. Jacobs v. Theimer. *Texas Supreme Court Journal* 18:222, 1975.

6. Reilly P: Genetic screening legislation. *Adv Hum Genet* 5:319, 1975.

7. Robinson J, Tennes K, Robinson A: Amniocentesis: Its impact on mothers and infants. *Clin Genet* 8:97, 1975.

8. Research group on ethical, social, and legal issues in genetic counseling and genetic engineering of the Institute of Society, Ethics and the Life Sciences: Ethical and social issues in screening for genetic disease. *New Engl J Med* 286:1129, 1972.

9. Komrower GM: The philosophy and practice of screening for inherited diseases. *Pediatrics* 53:182, 1974.

10. Levy H: Genetic screening for inborn errors of metabolism. Dept of HEW publication (HSA) 75-5708, Washington, DC, 1975.

11. Childs B: Genetic screening. *Ann Rev Genet* 9:67, 1975.

12. Lamberg L: Genetic screening: Learning what you never wanted to know. *Today's Health,* page 36, March 1976.

13. Beck E, Blaichman S, Scriver CR, et al: Advocacy and compliance in genetic screening. *New Engl J Med* 291:1166, 1974.

14. Pearson HA, & O'Brien RT: Sickle cell testing programs. *J Pediatr* 81:1201, 1972.

15. Powledge TM: Laws in question: Confusion over sickle cell screening. *Hastings Center Rept,* December, 1972, pp 3–4.

16. Hamerton JL: Human population cytogenetics: Dilemmas and problems. *Am J Hum Genet* 28:107, 1976.

17. Robinson A: Neonatal deaths and sex chromosome anomalies. *Lancet* 1:1223, 1974.

18. Conley R, Milunsky A: The economics of prenatal genetic diagnosis. Milunsky A (ed), in *The Prevention of Genetic Disease and Mental Retardation,* WB Saunders, Philadelphia, 1975, pp 442–455.

19. Hazard S, Carter FA: Preventing the birth of infants with Down's syndrome: A cost-benefits analysis. *Br Med J* 1:753, 1976.

20. Milunsky A, Atkins L: Prenatal diagnosis of genetic disorders. *JAMA* 230:232, 1974.

21. Schwarz RH: Amniocentesis. *Clin Obstet Gynec* 18:1, 1975.

22. Riccardi VM: Regional genetic counseling programs. Milunsky A (ed), *The Prevention of Genetic Disease and Mental Retardation*. WB Saunders, Philadelphia, 1975, pp 410–421.

23. Riccardi VM: Health care and disease prevention through genetic counseling: A regional approach. *Am J Publ Hlth* 66:268, 1976.

24. Parker WC: Some legal aspects of genetic counseling. *Prog Med Genet* 7:217, 1970.

25. Reilly PR: Genetic counseling and the law. *Houston Law Review* 12:640, 1975.

26. Rimoin DL: Manpower needs in human genetics. *Clin Res* 23:61, 1975.

27. Committee on Genetic Counseling. Amer Soc Hum Genet: Gentic counseling. *Am J Hum Genet* 27:240, 1975.

General References

A. Allen GE: Genetics, eugenics and class struggle. *Genetics* 79:29, 1975.

B. Beckwith J: Social and political uses of genetics in the United States: past and present. *Ann NY Acad Sci* 265:46, 1976.

C. Campbell AGM: Infants, children and informed consent. *Brit Med J II*: 334, 1974.

D. Davis BD: Genetic engineering: How great is the danger? *Science* 186:309, 1974.

E. Green HP: Genetic technology: Law and policy for the brave new world. *Indiana Law J* 28:559, 1973.

F. Hudock GA: Gene therapy and genetic engineering: Frankenstein is still a myth, but it should be reread periodically. *Indiana Law J* 48:533, 1973.

G. Lipkin M Jr, Rowley PT (eds): *Genetic Responsibility On Choosing Our Children's Genes*. New York, Plenum Press, 1974.

H. Milunsky A, Annas GJ (eds): *Genetics And the Law*. New York, Plenum Press, 1976.

I. Motulsky AG: Brave new world? Ethical issues raised by current approaches to prevention, treatment and research of genetic diseases. *Science* 185:653, 1974.

J. Neel JV: Genetics and society. *Genetics* 78:35, 1974.

K. Porter IH, Skalko RG (eds): *Heredity And Society*. New York, Academic Press, 1973.

L. Reilly P: The role of law in the prevention of genetic disease. in Milunsky A (ed), *The Prevention of Genetic Disease and Mental Retardation*. Philadelphia, WB Saunders, 1975, pp 422–441.

M. Tiefel HO: The cost of federal research: Ethical considerations. *New Engl J Med* 294:85, 1976.

N. Veatch RM: Ethical issues in genetics. *Prog Med Genet* 10:223, 1974.

CHAPTER SIXTEEN

THE COLORADO-WYOMING REGIONAL GENETIC COUNSELING PROGRAM

INCORPORATION OF GENETICS INTO ROUTINE PRIMARY CARE

Introduction

For fullest impact, the genetic component of primary care must be complemented by frequent local backup from genetic specialists. When the resources of a university-based genetics unit are coupled with the resources of local practitioners in the provision of health care, the net effect is much more than shifting services from one site to another (i.e., from medical center clinic to the local community). Rather, there is increase in the number of patients who receive genetic health care and an improvement of that care in individual cases. This has been our experience in the Colorado-Wyoming Regional Genetic Counseling Program (CW-RGCP) throughout Colorado, Wyoming, western Nebraska, and eastern Utah.

A Three-Part Approach to Genetic Health Care

The details of the organization and operation of the CW-RGCP, which began functioning in January 1973, have been reported in detail elsewhere (1,2,3). Outreach programs were in progress at this time in other areas, including northern and central California (4), eastern Pennsylvania and adjacent New Jersey (5), New England, and elsewhere (6). In the past two years there has been a proliferation of regional genetics programs in many different states. The general goals have been similar—to enhance genetic health care in local communities distant from major medical centers. The CW-RGCP, however, has been unique among outreach programs in two ways: 1. The large area covered (approximately 225,000 square miles) and relatively low population densities in most sections. 2. The three-part approach to enhancing local genetic health care, with equal emphasis on the provision of *clinical services, professional training,* and *public education.* In our efforts to improve patient care and develop a basis for prevention, we assumed that to develop local clinics alone would be self-limited. Thus, we made a concerted effort to start training programs for local and regional health care workers and educational programs for the general public.

By the end of August 1975, 13 clinics were operating, convening monthly or bimonthly. A total of 772 families had been seen, and 804 disorders diagnosed (Table 2-3, p. 15). Private physicians and nurses in the communities served as clinic coordinators and local sources of genetic information between clinics. In addition, each community contributed space and facilities for operating the clinics and maintaining patient records. The necessary secretarial work and the processing of genetic laboratory analyses was carried out at the Genetics Unit of the University of Colorado Medical Center in Denver. This meant that the local contributions were simple and inexpensive, key elements in initiating and maintaining clinics. Financial support for the program has been provided by the Colorado-Wyoming Regional Medical Program, the Henry J. Kaiser Family Foundation, the National Foundation—March of Dimes, the Wyoming Department of Medical and Social Services, the Colorado State Health Department, and the Department of Biophysics and Genetics, University of Colorado Medical Center. Annual operating budgets of about $80,000 cover expenses for personnel, travel, communications, and supplies. Minimal salaried personnel include a full-time medical geneticist, a nurse-geneticist or genetic associate, a cytogenetics technician, and a secretary. Additional medical geneticists serving on a voluntary basis are also necessary. Peer review of all cases at biweekly conferences has been critical in formulating a final disposition for each family. The patient care provided by the program has been as sophisticated and effective as that provided at the medical center, with the added advantage of being accessible to a larger and less financially able patient population. As explained elsewhere (1), the patients in the program have not merely been displaced from one clinic setting to another; most would not have received genetic counseling otherwise.

Training and Education

Regularly scheduled training programs are aimed mainly at physicians and nurses, although other health care workers and social workers frequently attend. Training efforts differ from "educational" efforts in their emphasis on techniques for implementation, especially history-taking, physical diagnosis, referral, and followup. At each clinic a presentation is usually arranged for physicians or nurses. Thus, recurring meetings with members of the health care profession in each community have become a feature of the program. In addition, special didactic presentations are periodically set up for in-depth instruction, sometimes coinciding with local events such as state medical society meetings. Accredited courses for physician and nursing education have been provided. Twice yearly there is a two-day workshop for public health nurses in the state of Wyoming. More intensive three-day training programs for selected nurses from each community have been held at the University of Colorado Medical Center. The net effect of these efforts has been to increase the genetic health care available to local communities beyond that provided by the clinics. This was demonstrated by: 1. Increasing numbers of referrals from physicians and public health nurses. 2. Increasing numbers of genetic cases handled by private physicians as indicated by an increase in telephone consultations with the physicians. 3. An increasing rate of referral for prenatal diagnosis. 4. Improved followup by public health nurses as indicated by increasing numbers of return visits and contacts with other family members. Primary-care practitioners were the physicians who most frequently utilized CW-RGCP services and who claimed to be most influenced by its activities (1,7). In the states of Colorado and Wyoming, we were able to project that there were only 1334 persons aged 20–44 for each general/family practitioner, obstetrician, or pediatrician who claimed to be positively influenced in patient care by the CW-RGCP (1). This relatively low ratio suggests that programs such as the CW-RGCP can have an effect on genetic health care over large areas.

The educational programs utilized public lectures; junior high school, high school, and college classroom presentations; magazine articles, newspaper stories, plus local television and radio appearances. In one instance an hour-long lecture and discussion was videotaped for repeated broadcast on the local educational television channel (Casper, Wyoming). The educational efforts had two purposes: First, *to make the clinical services maximally available* so that those who were aware of the services could use them without inordinate travel or expense. Second, *to raise the general level of understanding about genetics* so that more people would be aware of the uses of genetic counseling, in order that the counseling, if utilized, might be more meaningful.

We judged the educational efforts reasonably successful by two criteria; 1. The increasing numbers of self-referrals and referrals instigated by the patients or families. 2. The results of a random survey of local citizens in a community served by the CW-RGCP.

The purpose of the survey was twofold; to determine what portion of the population knew of the program and its availability, and to determine how many of these persons felt they needed genetic counseling. The questions asked are listed in Table 16-1. The community surveyed was Colorado Springs, Colorado, selected because of its similarity to many other American communities in term of size and geographic relationship to a major medical center. Colorado Springs had been involved with the program from the outset, 31 months before the time of the survey. The survey findings are given in Table 16-2. The conclusions we drew from these findings were as follows: 1. In Colorado Springs, 6.2% of the sample felt they could profit from genetic counseling. Projected to the entire community, this percentage indicates that about 3100 people between 20 and 44 years of age would want counseling. Even if we assume that 50% could be counseled by their primary physicians, 1550 persons in that age group would still require genetic consultation. Therefore, from the consumer's viewpoint, a need for a local genetics program is clearly present. The number of persons in the 20–44 age range who had already been counseled by the time of the survey (328 persons or 164 couples) represented 21.2% of the total (328/1550). Therefore, the CW-RGCP has taken a step toward meeting the needs indicated by the consumers. 2. Although only 9.9% of all the people interviewed were aware of the CW-RGCP, 47.7% of those who thought they could profit by counseling were aware of it. These findings suggest that the public education efforts may have contributed to a recognition of the need for genetic counseling.

Lessons Learned

Our experience in the CW-RGCP has taught us a number of lessons about the role of genetics in health care and the organization of regional outreach programs. These will be discussed in

TABLE 16-1

Questions Asked in a CW-RGCP General Public Survey, Colorado Springs, Colorado, July 1975[1]

1. Were you aware of the Regional Genetic Counseling Program?
2. Were you aware of the genetic counseling services it provides in your community?
3. Have you ever thought genetic counseling could be appropriate for you or any member of your family?

[1]Persons of stated ages 20–44 were interviewed in Colorado Springs, Colorado by asking these questions in the order given. Interview sites included three shopping areas preselected on the basis of varied socioeconomic levels of the respective neighborhoods. Intensive canvasing of men during lunch hour was meant to offset the preponderance of women interviewed during the rest of the day.

TABLE 16-2
Response to the Survey Questions of Table 16-1

	No.	%
1. Aware of RGCP?	139	9.9
2. Aware of community services?	50	3.6
3. Genetic counseling appropriate?	88	6.3
Total surveyed	1405	100

For the population between 20 and 44 years (1970 U.S. census) the survey sample represented 2.8%. Females made up 54% of the sample. (Data compiled by Mr. Greg Rosen.) The total Colorado Springs 1970 U.S. Census population of 135,060, includes 50,023 aged 20–44. Colorado Springs is about 75 miles from Denver near the foot of Pike's Peak. Of the 139 persons aware of the RGCP 46 (33.1%) estimated a need for its services, and conversely, of the 88 persons expressing the need 42 (47.7%) were also aware of the program.

terms of the program's interaction with local professionals and patients, and in terms of future planning and evaluation.

Interaction with physicians. Many physicians in private practice have an incomplete understanding of the role of genetics in modern health care. Thus, programs reaching into their communities must not only provide clinical services, but also persuade the physicians that there is a need for them. The support and enthusiasm of local practitioners help develop a program that does more than merely displace patients from a tertiary-care medical center setting to the local community. A fixed schedule of relatively frequent clinics (at least once every two months) coupled with training efforts is critical. If clinics are less frequent or are often cancelled, the program's local momentum is rapidly lost. General practitioners, family practitioners, obstetricians, and pediatricians are the main source of referrals (1,7). With their cooperation it is possible to enhance the management of most congenital and genetic disorders and reduce the need for referral to a tertiary-care center.

To facilitate this cooperation we began to implement the three basic notions elaborated in Chapters 1 and 2. The concept of *the genetic approach to human disease* as distinct from genetic counseling emphasizes the role and responsibilities of primary-care physicians in genetic health care, both for direct management and prevention. The *genetic classification of human disease,* a scheme for deciding whether a given disorder is genetic or calls for genetic evaluation, eases decisionmaking about referrals. Finally, the idea of the *family unit as the focus of health care* has emphasized the role of genetics beyond assistance to the effected individual. Additional cooperative efforts involve the ready availability of the geneticist to the local physician by telephone with encouragement to them to recognize and handle genetic cases on their own.

Interaction with nurses. Nursery, obstetric, and pediatric hospital nurses, office nurses, and public health nurses can play an important role in genetic counseling through direct referrals, patient followup, and liaisons with referring physicians. The public health nurse is especially important as a source of referral and followup for many patients who do not have a regular or smooth relationship with a physician, such as low-income city dwellers and relatively isolated rural families. CW-RGCP training efforts for nurses directed at recognition of genetic and congenital disorders, pedigree construction, and followup counseling have increased their contributions to genetic health care.

Interaction with patient and family. Involvement with the family and patient is optimal if the following considerations are kept in mind: 1. Self-referral must be allowed for. 2. Direct involvement with the family can take up to 4 hours or more and total time expenditure on behalf of each family averages 7½ hours (not counting travel time). 3. Many families do not travel to a tertiary-care center for genetic counseling (1) because the purpose of counseling is unappreciated or they could not afford it (this underscores the need for education programs and financial subsidy). 4. The team approach, with multiple levels of professionals, is well accepted by the families and allows for larger numbers of families to be seen. 5. Although the emphasis is on primary prevention, counseling also involves enhanced management and parental support, particularly efforts to assuage parents' sense of guilt, which in addition to having a positive effect on patient and family health can expand their utilization of community resources (secondary and tertiary prevention).

As pointed out in Chapter 2, CW-RGCP patients have varied and often complicated or rare disorders. For this reason the genetic counselor often becomes involved in patient evaluation and management, usually under the auspices of the primary physician. The overall care thus provided has been as comprehensive as in a tertiary-care medical center. It has benefited from conferences convened to discuss the outreach patients and ensure a consensus diagnosis and disposition. Photographing patients for these review conferences is a virtual necessity. Patient responses to this care have been assessed and found to be almost uniformly positive and frequently instrumental in subsequent decisionmaking about family planning (1,7).

Regional program planning. The key elements for long-term planning include: 1. The need for a three-part approach of providing services, training, and education. 2. Appreciation of the public health nature of genetic and congenital disorders. 3. Recognition of the special importance of prenatal diagnosis in preventing genetic disorders over widespread areas. 4. The need to organize the program with sufficient personnel to allow for frequent clinics and time expenditures averaging 7½ hours per family. 5. The need to keep local community obligations as simple and inexpensive as possible. 6. The need for financial subsidy, especially for chromosome analyses (2,3). 7. Recognition of the excellent training experience the clinics provide for clinical fellows, house-staff, medical students, and genetics graduate students. It is the public health nature of genetic and congenital disorders that establishes the logic of state financial support and the rationale for widespread and uniformly available services. By minimizing the occurrence and recurrence of genetic disorders through primary prevention efforts of all types, regional genetic outreach programs will easily justify the state's financial sponsorship in the long run.

Evaluation. The value of an outreach program must, of course, be judged in terms of the whole rather than the separate parts. It is not the techniques of genetic counseling or the counseling itself that are the consideration, but the effectiveness of the program to provide the counseling. Does it actually decrease the birth of individuals with severe disorders enough to offset the program's cost (8)? It is important, however, not to exact a cost–benefit analysis prematurely. The inertia any new regional genetics program has to overcome is so great that it should be allowed at least two years, and preferably three, before a stringent cost–benefit analysis is made.

Genetic Associates and Nurse-Geneticists in Local Genetic Health Care

A logical extension of the roles played by the genetic associate and nurse-geneticist in the CW-RGCP would be to have them operate in given areas as first-level geneticists. Under the auspices of a local health department or a private group practice, they would screen patients for genetic disorders and assist local physicians in deciding whether a patient or family should have a more detailed genetic evaluation. They would elaborate the family history, gather other historical data, perform a preliminary physical examination, and order various genetic screening tests. Then they could help the outreach programs geneticist and primary physician establish a final disposition. In some instances even the counseling could be handled by these workers. Our experience in the CW-RGCP suggests that in the long-run this will be the only way to provide comprehensive genetic evaluations on a routine basis to populations spread over large areas. This approach could substantially increase the impact of regional genetic health care programs by extending the range of effectiveness of a relatively small number of physician geneticists (9).

Cited References

1. Riccardi VM: Health care and disease prevention through genetic counseling: A regional approach. *Am J Publ Hlth* 66:268, 1976.

2. Riccardi VM: Regional genetic counseling programs. In Milunsky A (ed), *The Prevention of Genetic Disease and Mental Retardation*. Philadelphia, WB Saunders, 1975, pp 410–421.

3. Riccardi VM, Robinson A: Preventive medicine through genetic counseling: A regional approach. *Prev Med* 4:126, 1975.

4. Epstein CJ, Erickson RP, Hall BD, et al: The center-satellite system for the wide scale distribution of genetic counseling services. *Am J Hum Genet* 27:322, 1975.

5. Jackson LG, Peris LA, Cowchock FS, et al: Center-satellite programs for genetic services: the Philadelphia experience. *Am J Hum Genet* 26:44a, 1974.

6. Wendt GG, Theile V: A pilot scheme for a genetic clinic. *Humangenetik* 21:145, 1974.

7. Riccardi VM: Community response to a regional genetic counseling program. In Lubs HA, de la Cruz F (eds), *Genetic Counseling*. Raven Press, 1977 (in press).

8. Hagard S, Carter FA: Preventing the birth of infants with Down's syndrome: A cost-benefit analysis. *Br Med J* 1:753, 1976.

9. Rimoin DL: Manpower needs in human genetics. *Clin Res* 23:61, 1975.

General References

A. Rimoin DL: The medical genetics clinic and community health. Birth Defects Original Article Series, vol. VI, No. 1, pp 67–75, 1970.

B. Marks JH, Richter ML: The genetic associate: A new health professional. *Am J Publ Hlth* 66:388, 1976.

INDEX